Kil[illegible]
Zone

To and for those who serve their cause and country.

"A victorious army first wins and then seeks battle; a defeated army first battles and then seeks victory."

—Sun-tzu from *The Art of War*

Killing Zone

A Professional's Guide to Preparing or Preventing Ambushes

Gary Stubblefield
Mark Monday

Paladin Press
Boulder, Colorado

Killing Zone:
A Professional's Guide to Preparing or Preventing Ambushes
by Gary Stubblefield and Mark Monday

ISBN 0-87364-786-6
Printed in the United States of America

Published by Paladin Press, a division of
Paladin Enterprises, Inc., P.O. Box 1307,
Boulder, Colorado 80306, USA.
(303) 443-7250

Direct inquiries and/or orders to the above address.

CONTENTS

ACKNOWLEDGMENTS

Only an author knows what a team effort a book really becomes. Only the writer can begin to say what he owes, whom he owes, and how much he owes. That is particularly true for this book—a project that started with the discovery that there was a lamentable paucity of material on ambushes.

It seemed there was a little information scattered here, a bit over there, something mentioned in this or that book. But there was no single comprehensive guide or manual covering one of the oldest techniques of warfare known to man. It quickly became apparent that much of the best material resided in the brains and memories of my friends. Since so much of the information had yet to be written down, it seemed appropriate to be an amanuensis. From those beginnings the manuscript grew.

As a writer, you eventually realize that a complete acknowledgment of assistance would run longer than the manuscript itself. A book, after all, is the product of a lifetime of contacts, experiences, learning, background, and knowledge. Somewhere in the process of writing the acknowledgments, you also discover that, in some cases, you really shouldn't be acknowledging people by name. Publishers get antsy. Editors get upset. You start to trim the acknowledgments section back to a reasonable size. It is then that the writer realizes just how many people had a hand in the project, and how much they are owed. Soon you decide you'll just say "thanks" to most people personally. As a result, there is a long list of people I have to see in person.

However, there are some people who must not be left out—people like A.J. and Ann Monday, Blaise and Bill Pope, Don Kalick, Therese Pike, Anne Seligman, and J.A. D'Apuzzo. Over the years people such as these spurred me to write in their own, unique ways. I think it is quite fair to add, however, that absolutely none of them would have chosen this as a preferred book topic.

Other people serve as role models. J. Bowyer Bell is a friend and author. A man with a puckish demeanor, a wicked wit, and an inimitable style, Bow has a rare ability to string and sequence words. Reading a book by Bow makes it easy to understand why, throughout most of recorded history, writing was the province of gods and their priests. There is melody and lilt in Bow's words, as well as worlds of uncommon common sense. Reading one of his books makes you believe the text could, indeed, have been engraved on tablets by some miraculous process. Bob Friedlander, Neil Livingstone, and Rex Applegate are other consummate authors whose work and encouragement have served me in good stead—in this as in other writings.

Zbigniew (Stan) Stankiewicz has long been involved with military ambush in all its forms. From Dale June, who guarded two presidents in good times and the worst of times, came a great deal of insight into the problems of preventing and countering ambushes of protected persons. Tony Scotti's insights regarding auto ambushes would make a book in themselves. Gen. Ed Meyer has been more than helpful to me over the years.

Among others whom I can talk about, Al Moser offered me the first glimpses of this interesting and age-old tactic. It was Al who set my feet on the decades-long path of studying insurgency. And Matthew E. proved to me that people can, indeed, be induced to walk wide-eyed into an ambush even when they know exactly where the danger lurks. Scott (Scotty) Reeves gave me thorough reality checks on this subject, and a critical reading of the manuscript. Of course, acknowledgment should be made of my co-author, Gary Stubblefield, without whom this book simply would not have been possible. Our editor, Donna DuVall, deserves thanks and kudos far beyond a few words in print. Give her a raise, Peder. And after you say that, what remains?

Well, you need to know that any errors of omission or commission are mine alone. And I will be happy to correct them in following editions, for this is designed to be a living document that will change with time and technology.

Mark Monday
Norfolk, VA

From the outset of my interest in military operations, as a young child reading about the Red Coats on the British side being ambushed by the tactics of the American militia, it bothered me that ambushes were not more documented in literature.

Ambushes simply make good military tactical sense. They were the most frequently used tactic by U.S. Navy SEALs in the jungles of Vietnam. I have always considered the use of cover and concealment to properly surprise the enemy in a kill zone one of the safest and most effective means of taking on the enemy. However, a good ambush must be planned so that the tactics are properly employed. For future warriors to better understand the tactics of an effective ambush, they must be documented. My training came from such professionals as George Hudak, Gary Gallagher, and Wade Puckett; warriors with unparalleled experience in this area. My own experience refined my perspective on the value of ambushes.

Gary Stubblefield
San Diego, CA

PREFACE

A successful ambusher never attacks unless he is assured of winning. The successful ambusher follows Sun-tzu's dictum that battles should always be won before the actual engagement begins.

Successful ambushes are based on the ambusher's possession of complete intelligence. Success stems from an ambusher's willingness to undertake detailed planning. Only when the battle has been "won" in these ways can the actual attack be executed—with surprise, shrewdness, and violent determination.

Ambushes can be and are conducted against virtually anything that moves. The most frequently attacked targets are rail and vehicular traffic, as well as men on foot and boats on waterways. The targets may be either military or civilian.

Ambushers use forces that are sufficiently strong to smother the enemy; they destroy the opponent by quick shock action. Commanding ground, cover, concealment, and camouflage are used to the utmost. But the key component of any ambush is surprise. No ambush is effective if surprise is lost. All of the other issues take a backseat to surprise.

Those are the things this manual is all about. They will be repeated over and over again, in different ways and in different contexts, but those are the basics. They apply to all ambushes, no matter what the target or technique.

For the sake of simplicity and clarity, much of the

emphasis in this book has been placed on attacks against vehicles or vehicle convoys. But the same principles apply to ambushes of boats, trains, planes (yes, aircraft), and even individuals walking along a street.

At the same time, the book is designed to take note of the "special cases" that make the subject of ambushes so convoluted and so interesting.

The principles, cautions, techniques, and lessons listed here have been contributed in large part by experienced operators—professionals in every sense. Much of the material has been contributed by friends who rummaged through their memory and their closets for little tidbits of data that somehow seem to have been lost to the rest of the world. Many U.S. government documents—all now declassified—have been consulted extensively, though every effort has been made to remove the bureaucratese. For instance, to say that "an ambush is a trap sprung on a moving enemy and is based on concentrated surprise fire from concealed positions" is accurate. It's also boring enough to make a confirmed insomniac snooze. Curing insomnia is not our intention.

Much of the material in this book is written as if it applies to either a military or a civilian ambush. The differentiation of civilian targets from military targets in the text is really a distinction without a difference, but is one that is so deeply ingrained in the thinking of many people that it artificially divides the subject and makes it difficult to see the subject of "ambush" as a whole. In fact, most of the lessons in this book, while skewed in the text to refer to one or the other, can be applied to either military or civilian ambushes. The reader, or "operator," should look at the text from his or her point of view—military or civilian, ambush or counterambush—to get the most value from the material.

Ambush is an age-old technique, one that has been proven so highly successful that it endures. But ambush is not the ultimate tactic, one against which there is no defense or countermeasure. Units, individuals, and operators can have confidence that they have the ability to overcome the initial advantage of an ambushing force. Through prompt and determined action they will defeat an attack; through precautions they can prevent one.

Ambushes, as a tactic, can be discouraged; as a technique in war and social interaction, the ambush cannot be completely eliminated. But individual ambushes can be defeated; in fact, the ambusher often defeats himself.

Field "operators" concerned about being attacked must realize that, if caught in an ambush, there are methods that will not only reduce the effectiveness of the particular ambush but will so discourage the ambushers that the incidence of ambushes will drop in the future.

The keys to counterambush operations are early detection, immediate and vigorous reflex-type counteraction, and *relentless and intelligent pursuit.* While U.S. Army and Marine Corps doctrines often seem to stress the words "relentless" and "pursuit," the key word is "intelligent." There is no easier way to blunder into an ambush than to crash along through jungle or dense brush, relentlessly pursuing a quarry who has just staged an ambush. All the quarry has to do is pause long enough to set a 10-second delay on a claymore and stick the mine into the ground. Those "relentless pursuers" will never move again.

Each operator, whether military or civilian, must develop an integrated system of dealing with ambushes. It has to be based on his own knowledge, understanding, and personality.

Much of what follows are guidelines and general principles. These guidelines allow for considerable judgment in their application. Some of what follows are specific items that can be easily integrated into the particular approach of an individual. But, ultimately, the successful use of the material here is dependent upon the maturity, judgment, and professional competence of the individual. This is merely a guide to facilitate an early application of these basic personal qualities. This book is designed to stir the reader's thought processes, not to force anyone's thinking into a mold.

Whether you're a military leader trying to carry out an ambush against the enemy or a civilian security officer interested in preventing or limiting the damage from a terrorist ambush, keep in mind the successful ambusher's key rules:

1. Always have a good plan providing for every course of action of the enemy.
2. Always select an unlikely ambush spot.
3. Always put around-the-clock surveillance on the objective area up to the time of the attack if that can be done without compromising the ambush team, something that may be easier said than done.
4. Always get as much information on the target as possible, no matter how sketchy and tentative the data may be.
5. Always rehearse the elements of the ambush force.
6. Always achieve close control through rehearsals and effective communications.
7. Always vary the ambush techniques and design so that there is no set or consistent pattern that the opponent can rely on—and guard against.
8. Always have patience. It may be necessary to occupy an ambush position well ahead of time; patience is necessary if secrecy and surprise are to be achieved.
9. Always have effective camouflage and enforce camouflage discipline; an effective ambush cannot be achieved if men, weapons, and equipment aren't properly concealed.
10. Always strike quickly to gain surprise.
11. Always withdraw quickly by a different route than the one used to get to the ambush area.
12. Always have all-around security.

If you have specific questions or information you'd like to contribute about ambushes, the authors would like to hear from you. They can be reached by the following means:

Mail—P.O. Box 120008, Chula Vista, CA 91912
Compuserve—73540 , 2167
Prodigy—HCCK61A
Internet—safestor@crash.cts.com

1

Ambushes

The Big Picture

THE BIGGER THEY ARE, THE HARDER THEY FALL

Lt. Col. Francis Smith could look around the remnants of his column and see that it, as well as his reputation in the British army, lay in shambles. It was the April 19, 1775. For Smith and his troops it was a day of disaster.

It hadn't started that way. Only hours before Smith had led a proud column out, swinging along a Massachusetts road. His men had routed the rebel scum at Lexington; they reprised their victory over the rebels at Concord. While Smith had not captured all the supplies of war materiel he had been ordered to seize, some of the gunpowder under the control of the Massachusetts Provincial Congress had fallen into his hands. Because of his actions that morning, it seemed, there would be less danger to the men of His Majesty's forces in the colonies. Smith and his men had done their duty as British soldiers; they had dispatched the rag-tag ruffians.

By nightfall all of that had changed. Some 273 of the proud men whom he had led from victory to victory in the morning lay dead, dying, or wounded. The silence of those killed in action screamed as loud as the groans of those being treated, without anesthetic, by the military surgeon.

Along the route of return to Boston town, Smith's column had been ambushed and sniped at repeatedly. It seemed like every tree and hillock on the way back from Concord held an ambusher's position. Smith had won the battles. But the American colonists, lying in ambuscades, had turned Smith's triumphant return into a rout.

Ambushes of opportunity—that's what later writers would call such a series of improvised attacks that turned the red coats of Smith's men crimson with their own blood. Smith probably called it ungentlemanly war, war waged by cowardly ruffians and savages. A man with the stiff upper lip normally credited to all British army officers, Smith was not the first commander to rail against bushwhackers and ambushers! He would be far from the last.

Ambush!

It's a word that sends chills down the backs of those who run powerful governments. It makes brave men quail, and lesser beings turn coward in the face of that word—if the reality doesn't make them dead ducks.

Ambushes have a devastating effect on civilians and tourists, as well. Terrorist ambush operations against tourists virtually shut down the travel industry in Egypt during 1994. Their campaign put thousands of people out of work and hamstrung the government.

Companies and businesses that lose too many top executives to ambush operations—or find that the cost of security is eating up their profits—tend to leave these areas for greener pastures. Ambushers in Algeria, following the military example of their fundamentalist brethren in Egypt, proved that.

Ambush was the tactic that battered and bruised Americans in Vietnam. In Afghanistan, the effectiveness of the ambush—*zasada*—led indirectly, but inexorably, to the destruction of the vaunted Soviet bloc. The casualties conventional forces suffered in both places resulted in the demoralization of the troops—American and Soviet. The world's largest and best-equipped military forces, forces that had technology that included the H-bomb, were unable to prevent ambushes. They were unable to stop the casualties. That damaged—no, ruined—morale at the home front as well. The effective use of ambushes against Americans and South Vietnamese in Vietnam, against Soviet soldiers in Afghanistan, and against United Nations (UN) forces in Somalia proved the multimillennial wisdom of Sun Tzu:

> *When you do battle, even if you are winning, if you continue for a long time it will dull your forces and blunt your edge. If you besiege a*

citadel, your strength will be exhausted. If you keep your armies out in the field for a long time, your supplies will be insufficient. When your forces are dulled, your edge is blunted, your strength is exhausted, and your supplies are gone, then others will take advantage of your debility and rise up. Then even if you have wise advisers you cannot make things turn well in the end. Therefore, I have heard of military operations that were clumsy but swift, but I have never seen one that was skillful and lasted a long time.

Ambushes preserve the personnel and assets of the ambushers. They make war last longer. And if the ambushers can make it last long enough, they will destroy the will of their opponents—even if they cannot crush the military capability of that enemy. That's winning through attrition. It may not be gentlemanly. Purists might call it cowardly. But it is smart. It is effective. It has millennia of proven success behind it.

The climate and terrain of South Vietnam lent themselves readily to the execution of successful ambushes. The climate is temperate, permitting ambushing units to remain in position for extended periods of time without undue physical discomfort. And ambushers did. Available evidence indicates that the Vietcong often occupied ambush positions for as long as 48 hours prior to contact. Terrain also lent itself to the conduct of successful VC ambush. Jungles offer almost perfect concealment at the ambush site and provide concealed routes to and from the selected location. Mountainous terrain causes foot soldiers to become tired and less alert to ambush possibilities, and inexpert drivers invariably "close up" a vehicular column on steep grades. The relatively restricted arteries of communication in Vietnam limited routes that could be used. Secondary roads were often closed by VC sabotage, forcing the use of routes that could be interdicted easily by VC ambush.

On the U.S. side, the U.S. Navy's SEALs and the army's Special Forces were particularly adept at the ambush tactic as well. Their success proved that practically every geographical area of South Vietnam favored the ambush tactic—no matter who employed it.

The Soviet military faced many of the same sorts of terrain and ambush-linked problems in Afghanistan. The weather there was different, but the cultural and social environment they fought in had similarities to Vietnam. Though good soldiers, they too were bled dry by pinprick surprise attacks. Ambush after ambush, year after year, their lifeblood was leached out of them. The phrase "Red Army" came to have a new meaning. The tears of troopers' mothers, fathers, sweethearts, wives, and friends flowed as freely as the blood. In time the emotional, economic, and social drain of death and destruction undermined Soviet society.

In Somalia, the ambush-planning ability of a single tribal warlord gutted the UN and American desire to continue efforts there. After facing fewer than six months of concerted ambushes, the UN quietly sued Mohammed Farah Aideed for peace. Military men from around the world were routed by a group of drug-high clan members. The clansmen were armed with only the basic weapons of modern war—automatic rifles, grenades, and a few crew-served weapons. The women's and children's only arms were two, their left and their right. But they used those arms to throw stones and their voices to hurl threats. The women and children used their bodies as cover for the ambushing gunmen. Somalian guerrillas—men, women, and children—prevailed against helicopter gunships and laser-guided weapons.

This is an age of refined violence, where stand-off weapons such as guided missiles and "smart" weapons seem to give war an impersonality that was not common until the twentieth century. But the ambush was, and remains, a personal encounter: warfare at close quarters. To borrow from the world of communications, the AK and the claymore are "hot" weapons; the H-bomb and the Stealth fighter are "cold" ones.

When used properly, the hot ones can defeat the cold ones. All the H-bombs in the U.S. and Soviet arsenals have proven to be impotent in ambushes; claymores are not. A man can use a begrimed M16 or an aged over-and-under shotgun in locations where the pilot of a super-sophisticated Stealth fighter can sometimes only lay down a boom carpet in frustration.

Ambush. It's a word that the British troops trudging through the lush valleys and over the roadways of Northern Ireland hate just as much as Peruvian forces in the highlands of the Andes. It is a tactic that has been used against businessmen in Germany by terrorists and against communist rebels in the Philippines by government troops. It's a tactic that is only a little younger than the hills that hide the ambushers. It's a tactic more flexible than the curving vines that conceal the ambusher from his target.

Although conventional military forces can and do employ the ambush against one another, the ambush has taken on a new importance as a key guerrilla and terrorist tactic. Civilians—bankers, government officials, corporate executives—are just as much a target of bushwhackers as soldiers who are attempting to pacify a "disturbed area."

We live in an age where the superpowers have made an awkward peace with each other. Still, the surrogate paramilitary forces they once supported and controlled are now finding themselves free to strike out when they can and as they will. The ambush is the strongest tactic of the relatively weak. For that reason, a thorough knowledge of ambushes has become a life-and-death matter.

Not all attacks take place along a deserted road—one

that seemingly goes from infinity in one direction to nowhere in the other. Many ambushes take place on busy downtown streets or along well-traveled suburban roads. The ambushers don't hide behind trees or hunker down under triple-canopy jungle with snakes for companions. No, they sit quietly in a crowd at a cafe or stand nonchalantly at a corner, as if waiting for a bus, all the while looking for their prey. They drive by on their cycles or creep up from the rear in a car, their guns drawn but hidden from the occupants of nearby cars. Many modern ambushers blend in with their background—but their camouflage may be a business suit rather than jungle-patterned cammies.

Ambushes have a long, though some would dispute honorable, history in the tactics of war. Ambushes are a "natural" part of human conflict. Sun-tzu, the acknowledged genius of the Oriental art of war, alluded to them in the fourth century B.C. as one of the bases of war: "There are only two kinds of charges in battles," he said, "the unorthodox surprise attack and the orthodox direct, but variations of the unorthodox and the orthodox are endless."

Raids—where the attacking forces find a stationary enemy—and ambushes—where the attackers lie in wait for the enemy to blunder along—constitute Sun-tzu's "unorthodox surprise attack."

The warrior sage understood his subject well. Ambushes come in an endless variety. All the variations have earned the respect of the truly knowledgeable, no matter what the period or part of the world.

In some cases that respect almost borders on awe. For instance, it was ambushes that beat Fulgencio Batista and put Fidel Castro in control in Cuba. After years of pinprick attacks by rebel forces, the Cuban military was demoralized by the audacity and success of Castro's and Che Guevara's ambushers. They were so disheartened that the army refused to fight Castro's rag-tag troopers when they marched down from the hills to take Havana.

The military fled instead. In a set-piece battle, even at that time, Batista's men could have routed the rebels. But the rebels had pricked the troops so often and so hard that their will and morale literally bled away. Their fighting ability had been leached out of them.

Ambushes, when successful, are unbelievably devastating not only because of the carnage they cause, but because they point out so clearly the vulnerabilities of everyone involved—and uninvolved. They frighten those far from the attack with the knowledge that "it could have been me." They demoralize any who see themselves as a potential target. Ambushes also point out that a command, control, and communication (C3) problem exists among those attacked. A successful ambush screams "failure" at the losers and their leaders. "The king has no clothes" after a successful ambush—and the world now knows one can occur at any time!

It is in the area of command and control, all too often, that the ambusher succeeds while the ambushed commander fails, sometimes fatally. Dead bodies and blood mark the spot where a flawed C3 system was stressed beyond endurance.

In point of fact, ambushes can be defeated. While almost any ambush is likely to cause casualties among the targets, victims can even turn the table on their attackers. In fact the ambush is one of the best weapons to turn against terrorists and guerrillas. The ambush is not a guerrilla weapon or a terrorist tactic, but rather a weapon that anyone can use effectively.

The successful operator has to think of ambushes in both the defensive and offensive sense. Take a look at some ambushes—some famous, some not so famous, some infamous—to find the common denominator. Reviewing just a few days of violence worldwide will point out how widespread and effective ambushes are.

- On June 5, 1993, what was expected to be simply a

hot, routine day of drudgery for United Nations soldiers ended as the date of one of the bloodiest massacres in the history of UN peacekeeping forces. UN troops were planning on inspecting an arms depot of a local warlord; senior officers had invited the warlord to be present at the inspection. He didn't show up, but his "army" did—surreptitiously. Dozens of Somali gunmen, apparently keeping in contact by walkie-talkie, created a killing zone out of a stretch of road on the edge of Mogadishu. They pinned down a company of Pakistani soldiers and held off reinforcements. Nearly 100 Pakistani and U.S. soldiers were trapped by Somali gunmen for several hours by the heavy fighting in the streets of Mogadishu. They were eventually rescued by an Italian armored column. A simultaneous attack on troops guarding a food-distribution center whittled the ranks there. When the firefights were finished after more than five hours, a couple of dozen Pakistani soldiers were dead or mortally wounded.

• On June 7, 1993, a fundamentalist gunman and a gendarme were killed in eastern Algeria when the militants ambushed a patrol escorting a busload of prisoners to court. A pair of gendarmes and three of the 21 prisoners were wounded in the firefight in Tizi Ouzou province. None of the prisoners escaped.

• On June 2, 1993, in Turkey, Kurdish guerrillas attacked along the main road between Bingol and the provincial capital, Diyarbakir. About 50 PKK guerrillas stopped two passenger buses, ordered the passengers out, and set fire to the buses. There were no casualties.

• On May 26, 1993, a trio of Pakistani police officers and two bandits were killed in a gun battle. The local bandits—known as Dacoits—had ambushed a police van. During the encounter, which took place in hills along the border between the provinces of Sind and Baluchistan, the ambushers fired rocket-propelled grenades at the police van.

• On May 11, 1993, suspected Muslim extremists

bombed a railroad passenger car and a station platform in Manila; 23 people were injured. The attacks against the overhead railway transit network linking Manila and nearby suburban cities occurred as the army continued an assault on the hideouts of the Muslim fundamentalist Abbu Sayyaf group on southern Basilan Island. The railway attacks were designed to draw the government's attention—and force the armed forces—back to Manila. The blasts were caused by grenades linked to timing devices. They damaged three train coaches, but, more importantly, they disrupted one of greater Manila's most popular modes of transportation. One of the grenades, concealed in a plastic bag filled with fruit, exploded inside a train carrying hundreds of homeward-bound commuters. The other blast occurred minutes later at a waiting platform for train passengers in the Manila suburb of Pasay City. The blasts were timed to coincide with the evening rush hour when the commuter train was packed with passengers. That same day Communist rebels wounded at least 13 government soldiers during an ambush in the southern Philippines. About 40 members of the communist New People's Army mined the path of government troops on a resupply mission near Salay in Misamis Oriental Province, 490 miles south of Manila. Three soldiers were wounded in the mine blasts. The rebels then opened fire on the rest of the troops, wounding at least 10 others. The government dispatched troops, backed by armored personnel carriers and helicopter gunships, to track down the rebels.

What an array of actions—and it hardly begins to scratch the surface.

AMBUSH—THE BASICS

So what is an ambush? In the simplest terms, an ambush is a surprise attack upon a moving or temporarily halted enemy with the mission of destroying or capturing the enemy force. Ambushes are designed to harass and demoralize the enemy, delay or block movement of supplies, to free or take prisoners, acquire arms or supplies, kill selected persons, and channel enemy movement by making some routes unusable. Usually brief encounters, they do not entail the taking or holding of territory.

In an ambush, the target generally sets the time; the attacker sets the place.

The common elements of all ambushes are simple: the target is moving, or is stopped somewhere while in the process of moving, when it is attacked. The ambush differs in fundamental ways from the raid, though the two are sometimes confused. (A raid's purposes include destruction and damage of vital enemy installations, equipment, and supplies; the capture of supplies, equipment, or key enemy personnel; diversion of enemy troops from other operations; and the release of friendly prisoners.)

In an ambush, the trap is "sprung" when high volumes of fire or explosives are delivered into a killing zone—optimally a stretch of cleared area between 50 and 100 yards long.

The fire and explosives can be aimed at individual targets—a point—or they can be distributed over an area, with each ambusher responsible for filling a certain designated area with bullets and shrapnel. When the firepower is used against an area, rather than against a point, the zones of responsibility generally overlap by at least 25 percent on each side. There should be no "holidays" or gaps. Every square foot of the killing zone is covered—sometimes twice.

Ambushes are short, intense actions followed by complete and rapid withdrawal. The ambush is not designed to last over extended periods. The key to successful ambushes is shock action—the quick kill followed by equally rapid withdrawal upon completion of the mission. No attempt is made to hold the ambush site for extended periods of time. Normally, the greatest damage in any ambush is completed in the initial two or three minutes. What follows is only mop-up and the completion of any specific mission other than killing and destruction.

The rapid withdrawal from the ambush site is essential. In larger ambushes it is not unusual to find guides stationed at rally points immediately behind the ambush site; these guides direct members of the ambush party to safe areas. They thwart attempts by either the ambushed party or reinforcements to pursue. The need for guides is particularly keen when ambushes are conducted in darkness.

Guerrilla forces find ambushes useful because no ground is seized or held in an ambush. The ambush allows small forces with limited amounts of equipment and arms to harass or even destroy larger and better-equipped military units.

There is a tendency among some to see the ambush only as a weapon of the guerrilla, the terrorist, or the criminal. Officers and gentlemen all too often find an ambush to be an ungentlemanly way of fighting; they often forget they can use it to good effect themselves in counterguerrilla operations. This skewed, myopic view is often shared by police and the public.

The ambush should be considered a basic technique in counterguerrilla warfare. The ambush was, for instance, a primary technique used by the British in fighting guerrillas in Malaya.

The use of ambushes in counterguerrilla operations should not be considered a defensive tactic. Nor are ambushes necessarily the weapon of the weak. When properly planned and aggressively employed, ambushes represent an effective offensive operational means of defeating enemy forces and limiting their freedom of movement.

The effect of a successful ambush program is not measured merely by numbers of casualties, particularly in the counterinsurgency environment. From a counterinsurgency perspective, denial or restriction of freedom of movement, both during the day and night, is a most important benefit from an ambush program because the guerrilla unit must be able to move in order to live. (The successful guerrilla, on the other hand, can employ the ambush to demoralize regular forces and—most importantly—prove to the population that the government really is not in control. Again, casualties don't *necessarily* count.)

The continuous harassment, restriction of movement, and inability to acquire supplies that results from a government's ambush program will have an adverse effect on guerrilla morale and efficiency. It does on regular troops, as anyone who fought in 'Nam knows.

For the counterguerrilla, ambushes are effective tools because they force the guerrilla to engage in decisive combat at times and places that are unfavorable to the insurgent, they deny the insurgents their nearly invaluable freedom of movement, they deprive the rebels of weaponry and hardware that is hard to replace, they demoralize the insurgents, and they destroy or damage the infrastructure of hard-core personnel.

Ambush operations are dependent on current information about the location, movement pattern, and size of the opposing forces. In Vietnam, for instance, since the bulk of the VC movement was at night, most of the U.S./Vietnamese ambush operations were executed at night. Friendly units patrolled during the day and set ambushes at night. This same principle—that information is key—applies equally in the civilian realm, in "executive ambushes." In attacks against civilians, particularly protected persons, terrorists generally choose morning time frames and sites near the target's home or office to carry out their ambushes. One of the reasons is that these are the times and locations used consistently by the target. Executives are at their most predictable in the morning, as they leave for work, and it is easiest to observe and track them at that time.

The size of the ambush force to be employed and method of execution depend primarily on its purpose; i.e., whether the intent is to harass or destroy the enemy by use of a deliberate ambush, or whether the ambush is an attack of opportunity.

TYPES OF AMBUSHES

Ambushes, generally, can be broken down into two subsets: 1) *area/point ambushes* and 2) *deliberate ambushes/ambushes of opportunity.*

Area/Point Ambush

Area ambushes are used against enemy movement in a given area and are actually made up of a series of point ambushes. Point ambushes are established at the best locations to inflict damage on the opposition.

Deliberate Ambush/Ambush of Opportunity

Deliberate ambushes or ambushes of opportunity can be employed against either vehicular or personnel targets. Although both deliberate ambushes and ambushes of opportunity are conducted by military units, including irregular forces, the same is not true of terrorists. Terrorists seldom conduct ambushes of opportunity against civilian targets.

Deliberate Ambush

A deliberate ambush is one in which prior information about the target permits detailed planning before the ambush party leaves for the ambush site. In a deliberate ambush, the ambush unit is assigned a specific mission.

This type of attack is normally based on detailed intelligence. The intelligence includes size, composition, and organization of the enemy force, as well as the time the force will reach certain points or areas. When this information is not available, an area ambush may be established with several deliberate point ambushes located along the probable avenues of approach.

Also, stay-behind patrols can establish an area ambush by placing deliberate ambush positions on several objectives that have been previously cleared. In Vietnam, deliberate ambushes were also employed outside strategic hamlets for defense of the hamlet and to warn of an attack.

Keep this in mind: Particularly for terrorists, guerrillas, and other "irregular" forces that do not have large reserves of manpower, intelligence is fundamental to the planning of deliberate ambushes. Without accurate and timely intelligence, no rebel force can conduct an ambush effectively because, without reliable information, the risks to the ambushing unit are unacceptable by guerrilla standards. Required information includes knowledge of routes, composition, and time of arrival of the target to be ambushed; weapons and defensive troops that accompany the unit; and even knowledge of the unit commander—his capabilities and limitations. The state of training of the ambushing unit is always a factor; if it appears that an action cannot be successfully completed, it will generally not be carried out. Although this statement is couched in language that applies directly to military operations, it is equally applicable to ambushes against civilians and protected persons and to executive ambush situations carried out by terrorists. A potential terrorist target who makes it difficult for the terrorist band to get information on routes, composition of the personal security force, and depth of defense will probably displace the attack elsewhere. Terrorists will seek out another target, a person about whom the needed information is more readily available.

Deliberate ambushes can be planned against targets such as:

1. Any force or person about whom sufficient prior information is known

2. Units or targets that establish patterns by frequent use of the same routes or that habitually depart and reenter their home or base areas at the same points
3. Carrying parties or convoys that move at regular times over the same route; trains, aircraft, or ships that run on schedules.
4. Movements that form patterns, such as the changing of personnel on positions at regular times or the daily travel of an executive to and from his home.
5. Forces or individuals that are lured to a location through ruse or deception.

Ambush of Opportunity

An ambush of opportunity is one in which available information of enemy activity does not permit planning or establishing an ambush at a specific time, point, or in a particular area (see illustration on page 18). This type of ambush is normally employed when friendly forces see the enemy first and quickly establish an ambush to surprise and destroy him. Some experts call the ambush of opportunity a "hasty" ambush.

In some cases, patrols may be sent to an area, establish an ambush site, and attack the first profitable target that appears. The actual course of action is determined at the time when the opportunity for an ambush arises. To make ambushes of opportunity work, units must be trained thoroughly in the techniques of rapidly establishing ambush positions. Established standard operating procedures (SOPs) for setting up ambushes are extremely useful when an opportunity to carry out this type of attack arises.

Terrorists and small insurgent forces seldom, if ever, employ an ambush of opportunity.

AMBUSH MISSIONS

Many different missions are performed by ambush forces.

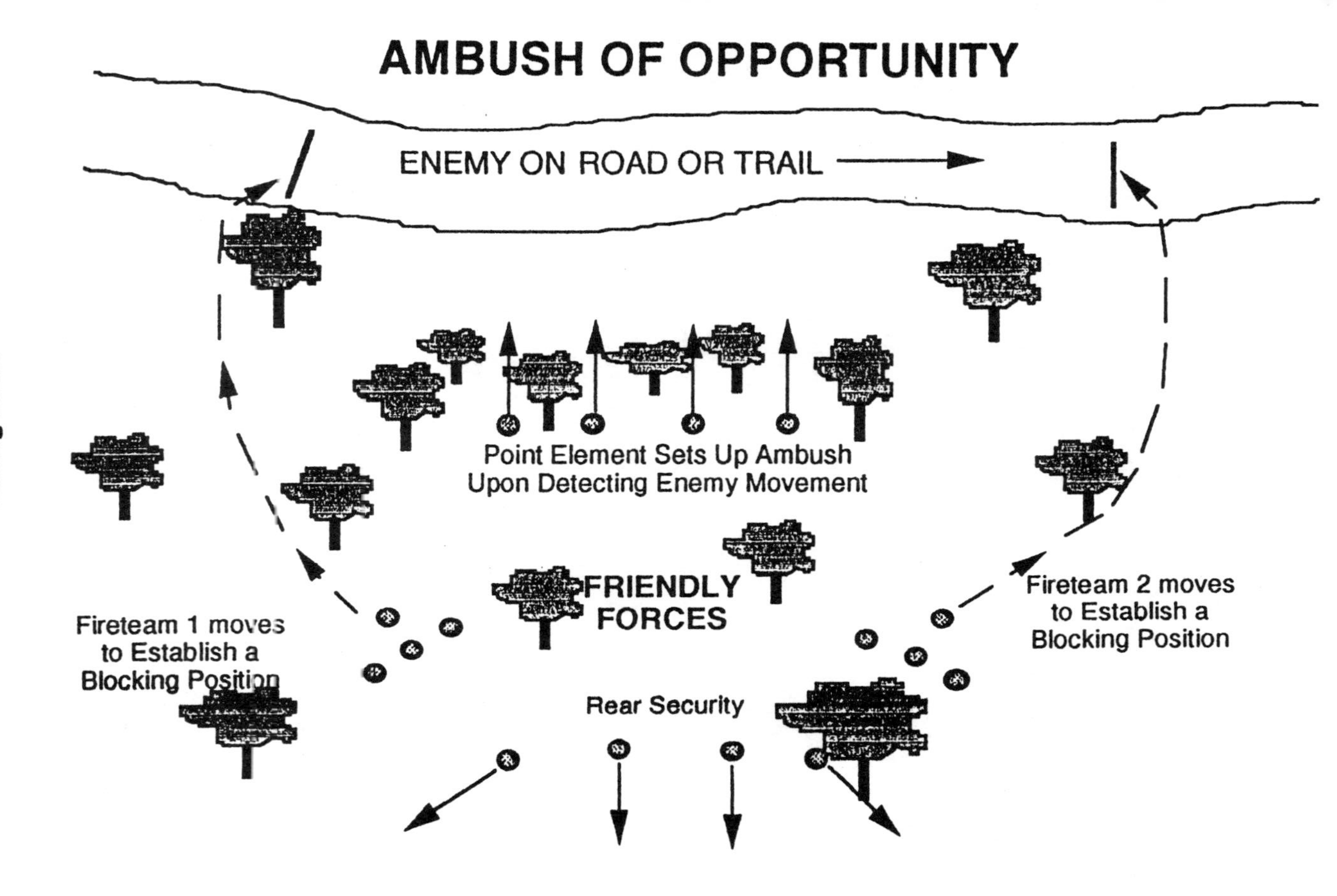
AMBUSH OF OPPORTUNITY
ENEMY ON ROAD OR TRAIL
Point Element Sets Up Ambush Upon Detecting Enemy Movement
FRIENDLY FORCES
Fireteam 1 moves to Establish a Blocking Position
Fireteam 2 moves to Establish a Blocking Position
Rear Security

An ambush may be defined by its purpose and the target against which it is directed.

Classes of Ambushes by Purpose

There are five different types of ambushes as defined by purpose:

1. *Harassing ambush.* This type of ambush is employed to harass and slow down the movement of personnel and vehicles of all types. No attempt is made to close with the enemy or destroy him. This type of ambush is frequently encountered during pursuit operations. It may range in size from a single rifleman/sniper to a relatively large-sized unit whose mission is to deny opposing forces freedom of movement. A harassing ambush can be overcome by bold, aggressive action because the ambushing force will normally avoid decisive combat. Executive ambushes of this type are virtually never found; terrorists find the cost and the effort too large for the expected result.

2. *Killing ambush.* This type of ambush is designed solely to kill personnel and destroy equipment. It is usually conducted by relatively large units, ones which may or may not close with the ambushed unit. The actions of the ambushing force will be governed by the composition and actions/reactions of the ambushed force. Counteraction in this type of ambush requires a high degree of training and determination because rapid and complete destruction of the ambushed party is the mission. This is a major type of terrorist ambush, directed at executives and other civilian-protected persons.

3. *Resupply ambush.* This type of ambush is frequently encountered in guerrilla warfare. It is designed to equip or re-equip, supply or resupply, guerrilla units—units that have inherent supply problems. Complete destruction of the ambushed party is not the mission of the ambushing force. Force must be tempered because overdestruction will defeat the purpose of the resupply mission.

Since the ambush's killing power must be well measured, counteraction or defensive measures are usually effective. The presence of "carrying parties" at the ambush site—to retrieve and remove the supplies or equipment—often means that there are large numbers of less well-trained personnel at the scene of action. This can betray the presence of the ambush party. Also, counteraction in a resupply ambush is more effective and pursuit easier; bearers are slowed by the burden of their loads. This type of ambush is seldom mounted against executives and other civilians.

4. *Prisoner ambush*. Put simply, this is a kidnap ambush. It is a difficult ambush to execute and probably the easiest to counter. Since the mission is to seize one or more prisoners, the ambush is not primarily designed to kill or destroy. A definite target is usually prescribed. In military ambushes, the target is often a VIP, a courier, an officer, or some specific individual. Sometimes the prisoner is to be held for ransom or bargaining purposes, sometimes for intensive interrogation. Since the mission is to take this individual alive, use of killing power must be limited and even avoided. Where possible, the targeted person must be isolated by gunfire and sometimes by movement in order to carry out the abduction. This results in a tricky balance, and firm counteraction on the part of those who are ambushed often upsets that balance, leading to the defeat of this kind of attack.

The prisoner ambush, often used by the military, is also frequently used in civilian attacks. In civilian life, the prisoner ambush (most people just refer to it as a kidnapping) is directed largely against officials, executives, or members of their families, by terrorists.

5. *Combination ambush*. One or more of the ambushes described above may be combined, providing that the missions are not completely incompatible. In Vietnam, ambushes were rarely executed for a single purpose. In the case of the Vietcong, for instance, their ambushes almost

invariably were designed to kill personnel and capture supplies from the ambushed party. It should be pointed out, however, that any combination of missions makes the counteraction easier because multipurpose ambushes tend to become more complicated in execution.

Classes of Ambushes by Target

Ambushes can also be classified by their intended targets. There are five categories of these:

1. *Ambushes against dismounted troops.* This type of ambush may be conducted against troops either in column on trails or roads, or in a deployed formation. This type of ambush is considered extremely difficult to conduct against military units because trained troops moving on foot should be alert for the possibility of ambushes. They usually have security elements preceding, following, and flanking the column.

Most examples of successful ambushes of this type involve targets that are small, poorly trained, and poorly equipped. However, large regular units have been ambushed and sustained heavy losses. Such ambushes, however, are the exception rather than the rule. Larger units are far more difficult to ambush with near-certainty of success (a prerequisite of any guerrilla operation) because they generally employ adequate security and are too large to be defeated in detail. Instances of successful ambushes of large units are *always* characterized by serious breaches of security and failure of the commander and his troops to take rapid, positive, counteraction.

Paradoxically, in the civilian world this is one of the easiest and most sure-fire types of executive ambushes (see number 6 below). It can usually be carried out easily where intelligence information shows the target can be found walking. Most civilians and executives are *not* wary while they're walking from place to place; they are not looking for trouble along the way. There are no security elements, in most cases, and the target cannot move far or fast. The executive caught in an ambush while walking has little hope of outrunning a bullet fired by a pursuing terrorist, even if he could run with the speed of a cheetah and the grace of a gazelle.

2. *Ambushes against wheeled or tracked vehicles.* This ambush is one of the easiest to conduct and one of the most difficult to counter, whether in the military or civilian arena. The speed with which vehicles move is no defense against ambush. Rather, in most cases, that speed becomes a liability that must be carefully considered. Experience indicates that although convoys often carry quantities of weapons capable of delivering a heavy volume of firepower, and even though there may be vehicles whose sole purpose is to protect the convoy, there is often a failure to provide adequate security at the point and along the flanks. Often there is an insufficient interval between vehicles.

Convoy organization, march control, and discipline are absolute requirements for secure motorized movements. When making motor moves, personnel must be assigned specific duties to perform during the move. Leaders must be appointed for each vehicle. Each individual must be aware of his responsibilities during the move and the actions to be taken if ambushed. When the personnel making the motorized move do not come from a single unit, they must be rehearsed in counterambush drills until they can execute the local SOP drill without hesitation.

Once a close-interval vehicular column is stopped by an ambush, the destruction of the entire convoy is relatively simple. Unprepared troops involved are generally surprised, grouped in easy targets for automatic weapons and grenades, and forced to dismount before they can effectively engage the ambush force. Only through the application of proper preventive measures and counteraction can the effectiveness of this type of ambush be reduced.

In executive ambush situations of this type, surprise is a major factor that gives terrorist attackers a great advantage. Terrorists are also aided by the targets' realization that any resistance on their part will likely result in the killing or injuring of passersby—whether the targeted driver uses the car as a weapon or the bodyguards trade gunfire with the attackers. The terrorist has already decided that the dangers to the uninvolved are not an issue; the target and the bodyguard or driver have to come to that same conclusion in a single moment. The instant of hesitation, brought about by concern for others, can be the difference between life and death.

3. *Ambushes of watercraft moving along inland waterways.* The ambush of boats and small craft moving along relatively narrow inland waterways is, in many respects, similar to the ambush of vehicular columns. The problems of countering the ambush are amplified by the fact that, prior to any effective counteraction, troops must first reach shore—often a difficult task. Further, putting out point and flank security along streams, rivers, and lakes is difficult.

This type of attack is seldom found in the civilian arena, although Muslim fundamentalists have used it with terrifying effect in Egypt to destroy the tourist industry. Although most of those attacks failed in every tactical sense, they resulted in a publicity coup and must be ranked as highly successful in the strategic sense.

4. *Ambushes of trains.* This often combines elements of sabotage and raid operations. Train ambushes are considered highly unusual in conventional military operations but are more common in guerrilla/terrorist operations. Some terrorist groups have a record of, and seem to specialize in, railroad ambushes.

5. *Ambushes of aircraft.* Landing zones or the ends of runways are the most vulnerable points for aircraft, which can often be attacked from outside the airport proper by anything from rifle fire to shoulder-fired surface-to-air missiles (SAMs). Aircraft on their final approach and landing, and those just taking off, are the most vulnerable.

This type of attack has shown a low rate of success, but until recently the technology to shoot down aircraft from the ground has been primitive. The proliferation of small, easily fired SAMs, particularly as a result of the war in Afghanistan, and their distribution to terrorists around the world, make this form of ambush a future nightmare.

6. *Ambushes of individuals.* This can be anything from an attack on a head of state or a key employee at a company to a person of the wrong political persuasion or ethnic background. They can be attacked while riding in vehicles or while walking. These types of ambushes, sometimes using cycle-mounted gunmen or remote-controlled roadside bombs, are difficult to fend off. They are best referred to as executive ambushes. The methodology often involves target types 1-5, as noted above.

You may run across other classifications. There are any number of schemes, and most have validity. But all classification systems are artificial. For instance, a government study conducted during the Vietnam War defined U.S. ambush missions as:

> *1. Capturing or destroying VC attack or raid forces in the vicinity of populated areas.*
>
> *2. Ambushes utilized as a defensive measure in protecting hamlets and villages.*
>
> *3. Capturing or destroying groups of VC as they attempt to leave or re-enter their war zones.*
>
> *4. Ambushes executed in order to kill VC leaders.*
>
> *5. Ambushes set by stay-behind forces in conjunction with tactical operations.*
>
> *6. In search and clear operations, ambushes are set to intercept the enemy being driven into the ambush position by the searching element.*

This ambush mission can be used in conjunction with the "fire flush" and "rabbit hunt" techniques of searching an area.

7. Ambushes conducted against targets of opportunity.

Yeah, things did get a little confused and fuzzy over there at times. And there is still a great deal of confusion.

COMPOSITION OF THE AMBUSH FORCE

Some ambushes can be conducted by a lone gunman, as in an assassination. But in the more formal military-style ambush there are often three parts to the ambush force: *a command element, an assault element*, and *a security element.*

The *command element* is made up of the commander, communications personnel, observers, medical team, and any other liaison personnel. The *assault element* captures or destroys the enemy and carries out the mission. The assault element is further broken down into assault teams, support teams, and special task teams.

The assault team accomplishes the primary mission: to kill or capture the enemy. The support team provides fire support for the assault element. This team is armed with a machine gun(s) and/or mortars and mines. The support team prevents the enemy from escaping through the front or rear of the killing zone. If a demolition team is to be employed, it is part of the support team. The special task teams eliminate sentinels, breach obstacles, destroy targets, lay mines, and conduct searches. The mission of the search party is a key one: to search the dead and wounded for documents and to pick up weapons and ammunition and equipment. People may be assigned solely to a special task team, but in most cases, the special task duties are assigned to individuals of the assault and support teams.

The *security element* provides all-around security. It protects the assault and support elements, and it covers all areas of approach into the ambush site that the enemy might use to reinforce the ambushed force. Its job is also to provide early warning of the arrival of the enemy force. This element also covers the withdrawal of the assault, acts as a rear-guard where necessary for the command and assault elements, and secures the rally point. In large ambushes, the security element is divided into separate teams.

The role of the security element is crucial. During the ambush itself, the ambushing force is extremely vulnerable to organized counteraction, especially flanking maneuvers that threaten escape routes. Once the ambush is sprung, the attention of the ambush party is totally directed at the accomplishment of its mission, and attention is therefore focused almost exclusively on the ambushed unit. Only by providing security forces along its flanks and rear can the ambush party protect itself from being surprised in an effective counteraction.

But all of this discussion about multiple elements tends to obscure a simple fact: *Ambush parties should be kept as small as practicable.* The tendency to make ambushes too large should be avoided. In Vietnam, for instance, five to eight men was considered a good size.

SECURITY OF THE AMBUSH SITE

A key characteristic of the ambush is complete security of the ambush site until the ambush itself is sprung. Ambushes are intended to be impossible to detect until the ambushed party is within the killing zone. That's the whole essence of an ambush. Extraordinary care is exercised by the guerrilla force, terrorist team, or military unit to maintain the ambush site in a state of readiness without revealing its presence—in other words, in a state of complete security. All-around security is strictly maintained; friendly civilians who stumble upon the ambush site are detained until after the ambush, and entrance or exit to the site is never along routes that can be seen by an approaching unit.

In rural areas, when mines are placed in roadways or along the "far" side of the road, access to the road is gained some distance from the ambush site or at a cross trail to preserve the natural appearance of the site.

The VC would sometimes completely prepare their positions, then withdraw the ambush party to a "safe area" nearby to await the force to be ambushed. Only lookouts remained in the ambush site, the main ambush party would occupy the attack site only at the last possible moment.

Other VC variations designed to improve the security included ambush sites that were completely underground and elaborately camouflaged. In some instances, the ambushers lay in wait under water, breathing through hollow reeds. This latter technique was used with considerable success in Vietnam against the French and Americans in the Capital Military District.

Techniques such as these ultimately contribute to the security of the ambush site and assist in gaining surprise when the ambush is sprung.

AMBUSH FORMATIONS

Typical ambush formations are linear or on-line, L-shape, V-shape, demolition, bait trap, and pinwheel types. These formations can be employed effectively as deliberate ambushes or ambushes of opportunity.

Most often the ambushing party will assume that the enemy or opposition force will be moving from a certain direction, going in another direction. In most cases this will be true. But a good ambush must be capable of accepting the approach of an enemy force from more than one direction.

Because of the limited size of the ambush force, the ambush commander may be able to execute an effective attack from only two directions. Nonetheless, the other directions must be covered by security elements who make certain the ambush forces are not attacked from a surprise direction and who provide early warning of any intruders from an unexpected direction.

Linear or On-Line Ambush

The on-line ambush, which uses fire from one flank to saturate the killing zone, provides the simplest example of the ability to accept an enemy force from multiple sides. It accepts contact from the front—where it is expected—the right, and the left. The security element at the rear secures the ambushers against surprise attack, but it is probably not strong enough to ambush an enemy force—only to engage it and give the ambush commander time to redeploy troops to meet the new threat and probably to withdraw. This ambush is perhaps the easiest to control. Communication within the ambush line is easiest, and the maximum firepower is concentrated straight forward, along one flank of the targeted individual or group.

The linear formation is often sited where it is impossible or difficult to move off the trail or road—for instance when there is a steep hill or embankment on the side away from ambushers. Mines and booby traps—including det-cord laid in ditches, hollows, and likely areas of cover and concealment on the far side of the ambush—are often used to keep the ambushed party on the road or trail and force them to remain in the central killing zone.

There are disadvantages in using linear ambushes. They are easily flanked and are difficult to use against large formations. Sometimes the line is positioned so that it is set perpendicular to the line of approach—a technique referred to as "crossing the T." In such a formation, the ambushers will engage the leading part of the enemy formation when it enters the killing zone. Generally this

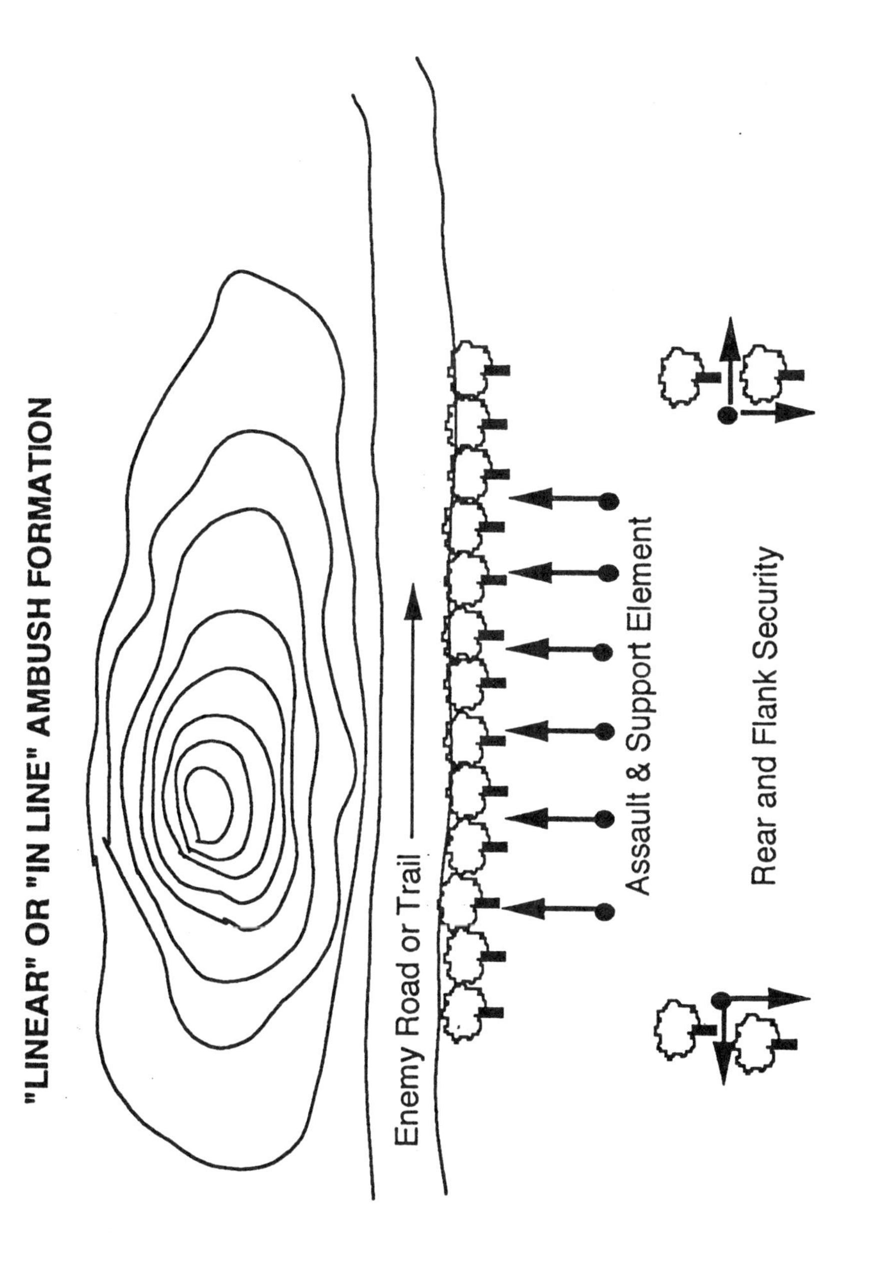

"LINEAR" OR "IN LINE" AMBUSH FORMATION
Enemy Road or Trail
Assault & Support Element
Rear and Flank Security

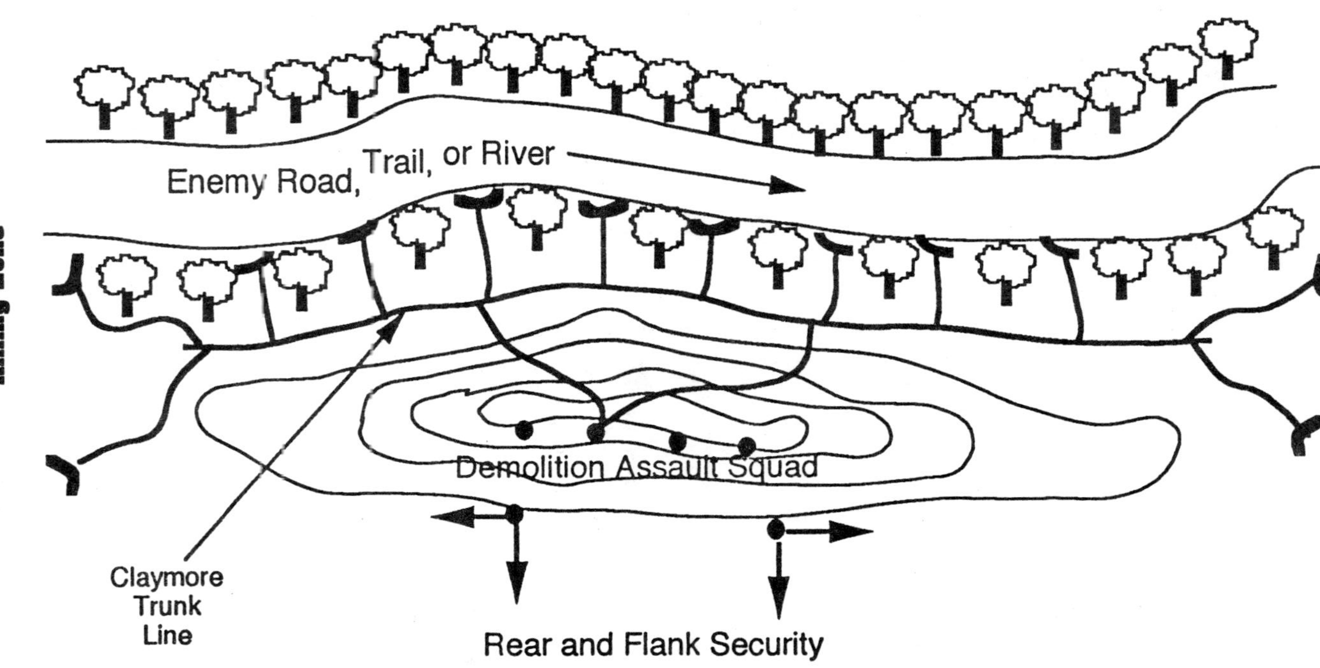
LINEAR CLAYMORE DEMOLITION AMBUSH FORMATION
Enemy Road, Trail, or River
Claymore
Trunk
Line
Demolition Assault Squad
Rear and Flank Security

variation of the linear ambush is not a favored method, and for good reason. It permits the enemy's rear elements to maneuver and gives them an opportunity to flank the ambush. However, it will work well against a small force that is not using a point man.

L-Shape Ambush

The L-shape ambush sets up fire from two directions. It is a natural and highly effective ambush along curves and bends, though it is somewhat more difficult for a commander to control effectively. It often lends itself well to ambushes along rivers and canals. Careful placement of the machine gunner and other heavy weapons is extremely important here.

When setting up an L-shape ambush, the long side is usually parallel to the expected route of the target. The ambush commander should generally deploy his forces so that the enemy entrance and exit are perpendicular to the short leg of the ambush. Claymore mines placed on the uncovered flanks are useful to prevent any flanking attack—or to make it costly in casualties. When stand-off weapons are used in an L-shape ambush, they should be placed close to the apex of the ambush, or the center of the ambush scrimmage line, to reduce the likelihood of friendly-fire casualties on the flanks. Automatic weapons set up in the middle of each leg, or close to the center of each leg, allow for greater overlapping fire in the killing zone. When automatic weapons are set up that way, the enemy is caught in what is effectively cross-fire.

L-shape ambushes lend themselves to a technique called a *staggered engagement*. In a staggered engagement, one leg of the ambush fires. From the target's position the attack appears, at first, to be a linear ambush. Then, at a pre-set time or in special circumstances, the second leg opens fire. The special circumstance could be something such as the initial leg's inability to suppress enemy return fire or efforts by the enemy to organize a counterattack.

Against convoys, a first leg using mortars, rockets, and heavy automatic weapons could attack the major vehicular targets. The second leg would then use suppressing automatic weapon and rifle fire against the enemy personnel as they detrucked.

V-Shape Ambush

V-shape ambushes are useful in some areas. When there is sufficient depth along the arms of the V, these have the advantage of denying the ambushed party a safe area in any direction. If the ambushed individuals concentrate fire toward their front, they can be killed from the rear; if they turn around to deal with the danger that is behind, they expose their backs to other dangers. The major problem with this type of ambush is that bullets and shrapnel that fail to find a billet among the ambushed can easily kill or wound some other member of the ambush party. When the legs of the V are long, the ambushers are effectively shooting in the direction of their own people. But when this form of ambush is employed from a height, so that the ambushers are looking down and shooting down at their targets, the rounds that miss go into the ground. Command and control problems are moderate, and, again, placement of the machine guns is important.

Demolition Ambush

The demolition ambush avoids, where possible, the use of guns. Explosives such as claymores, fragmentation grenades, or concussion grenades stitched together with det cord are often used in military attacks. Car bombs or roadside explosives are often used in civilian executive ambushes, as well as in some insurgent ambushes of military forces. The demolition ambush often involves radio-controlled explosions set off on command from a safe distance.

Demolition ambushes are extremely useful when engaging an enemy that is either numerically superior or has far heavier firepower.

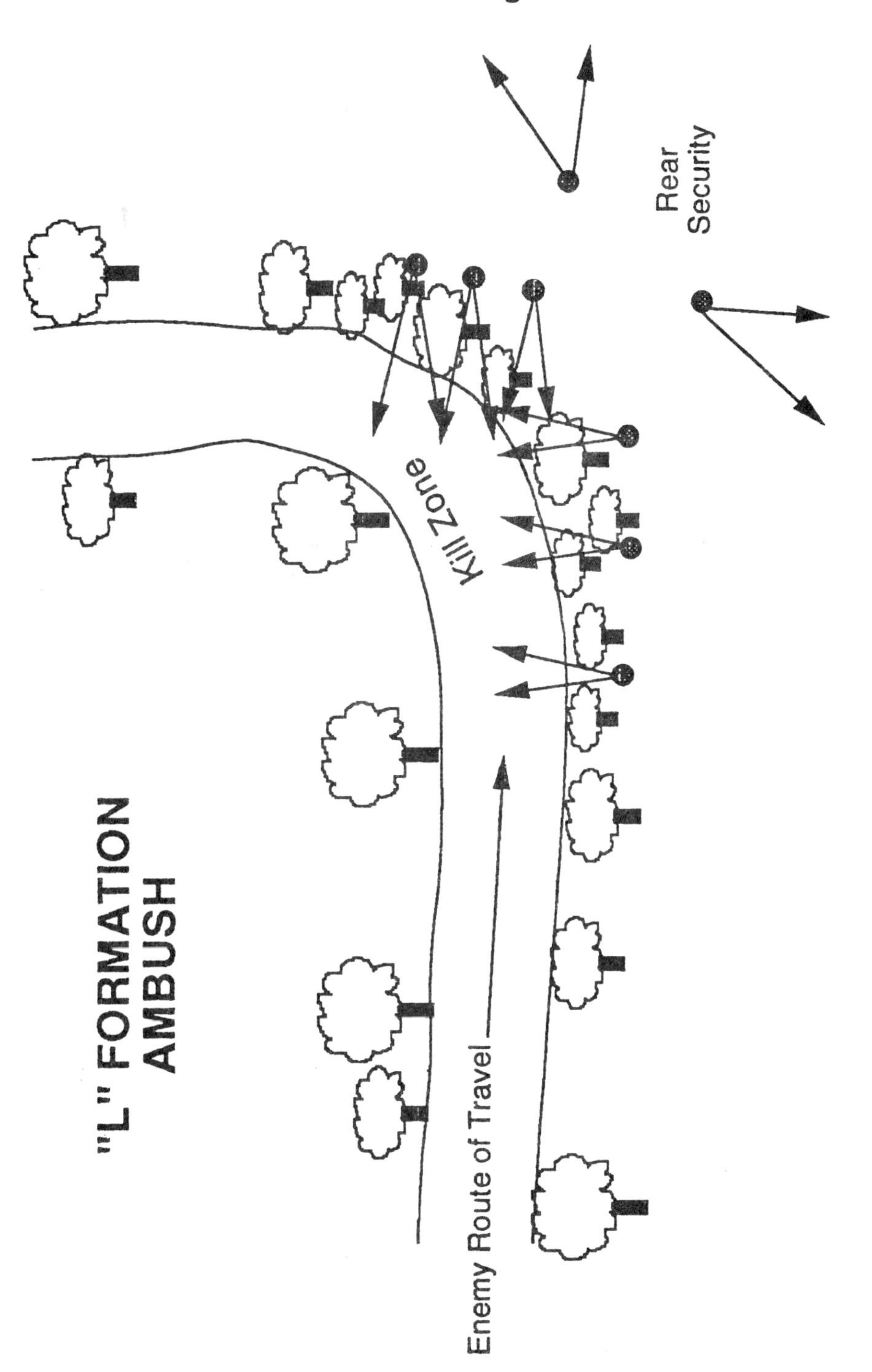
"L" FORMATION AMBUSH
Rear Security
Kill Zone
Enemy Route of Travel

"V" SHAPED FORMATION AMBUSH

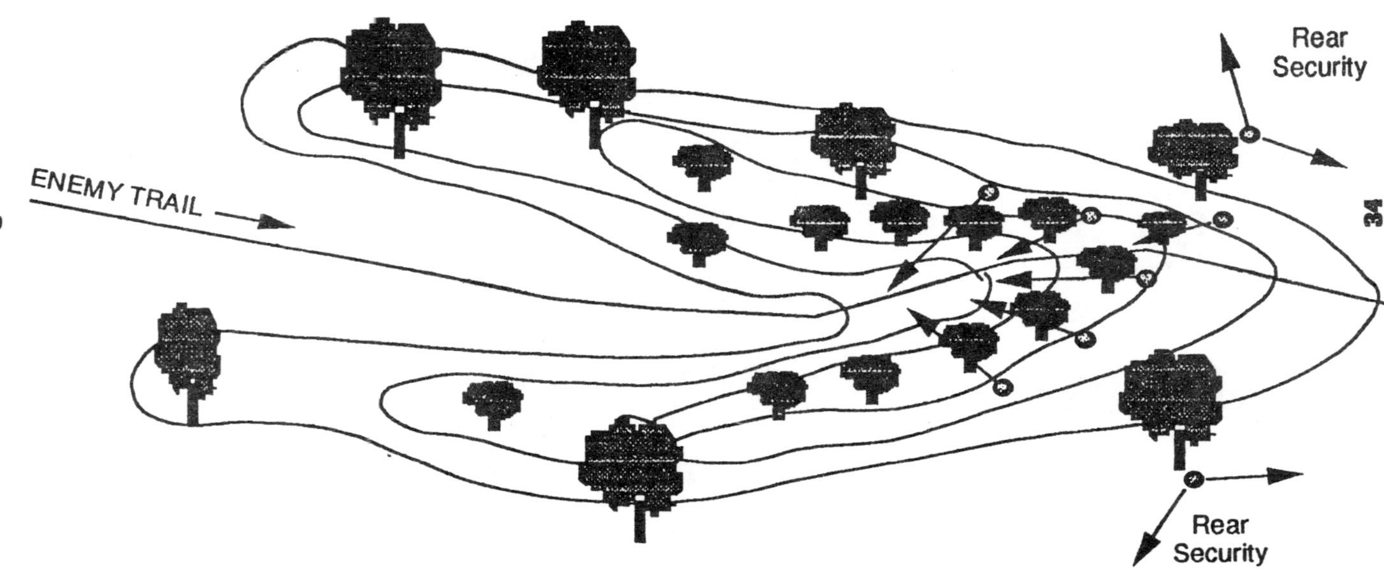

In a demolition ambush where no weapons are fired, the enemy might not be certain if he tripped a booby trap. At worst the target will tumble to the fact that he is in an ambush but almost certainly will be unsure of how many ambushers there are. The explosives should be set off by the commander at a safe observation post, i.e., one that is not within effective gunfire range if the ambush fails or is only partly successful.

One key to demolition ambushes is to use flankers with radios to advise the commander when the target is in the killing zone. Silenced (or more correctly, suppressed) weapons can be used by snipers to pick off members of the ambushed unit who escape the blast effects. The greatest drawback to this type of ambush is the time required to set it up.

Bait-Trap Ambush

The bait-trap ambush is useful when the ambusher knows the target's SOP and is certain of the action the target will take once the firing starts. If effect, the idea is to force the target into a killing zone through the use of his own tactics—for instance, opening fire on one flank when you know that will drive him toward the opposite flank, where you have set up a demolition ambush of carefully emplaced shaped charges and antipersonnel explosives. Some bait-trap ambushes can be set up by leaving a dead body or weapon alongside the road. Troops or police are sure to investigate and, if they don't investigate their own safety before investigating the bait, they walk open-eyed into the killing zone. Although not a formation in the conventional sense, this type of ambush starts with a bait, and the shape of the trap is tailored to the bait.

Pinwheel Ambush

Pinwheel ambushes are somewhat unusual and are generally used when it is not clear exactly from which direction the ambushed party will be coming. This type of ambush formation provides great flexibility in certain circumstances and has a high degree of rear and flank security. Pinwheel ambushes are often used at road or trail crossings where the ambush targets may be coming from any direction. A notable feature of pinwheel ambushes is that the support element is deployed in the center of the ambush position so that it can be effectively and quickly employed in all situations and in all directions.

PREPARING THE AMBUSH

The best ambushers are those who learn about their targets. They know how many people will be in the killing zone, they know how to recognize them, know if they use guides, what kind of formations they are likely to use, which direction they will be moving along which route, the kinds of weapons they will have, what times they will be on the move, and whether they will have point and rear-guard elements. In executive ambushes the terrorist will be able to recognize the individual, know who will be with him or her, what guards there are likely to be, whether any car involved is armored, whether there are follow cars, and whether the target can directly call police or other support.

The successful ambusher knows whether the target will be afoot, or what kind of transportation the target will be using and how that relates to speed. He or she knows what supporting forces will be available to answer any calls for help and what kinds of rapid communications are available to the ambush target to call in help.

The best ambushers study the terrain carefully to learn if there is a pattern of activity or natural funneling that will assist in setting up the ambush. For instance, does the target make a daily sweep of a section of road, one that has high banks on both sides?

Where possible, the ideal ambush sites should be avoid-

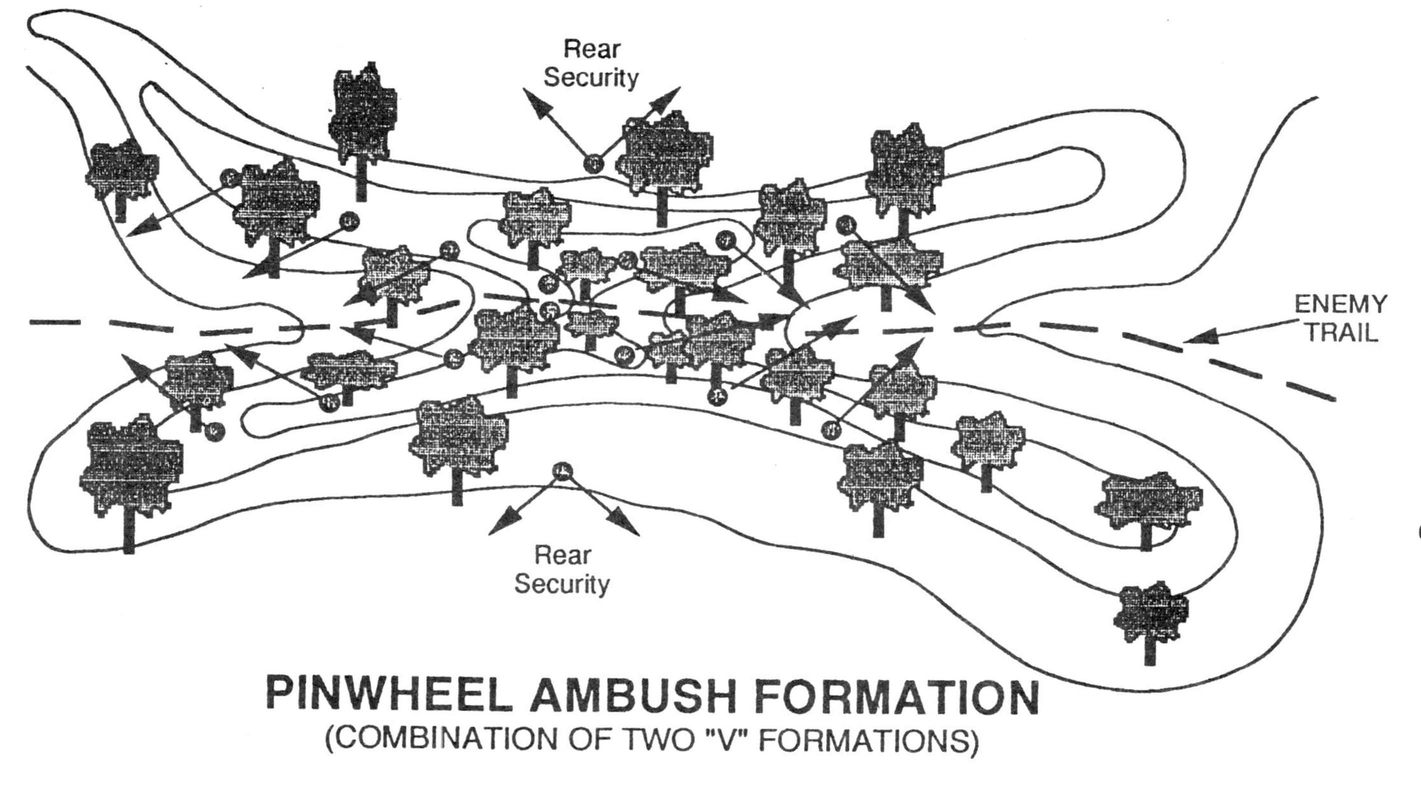

PINWHEEL AMBUSH FORMATION
(COMBINATION OF TWO "V" FORMATIONS)

ed. An alert enemy is always suspicious of ideal sites and goes around them where possible. Even if the enemy doesn't avoid such sites, he is more alert when in the vicinity of them; surprise is much more difficult to achieve.

Rather than seeking the ideal site, look for one that meets four criteria:

- Will channel the enemy into the killing zone
- Has favorable fields of fire
- Allows for preparation and occupation of concealed positions
- Has covered routes of rapid withdrawal

It helps to know the target's SOPs and counterambush procedures. It also is important to know whether troops and commanders are likely to use those SOPs and procedures. If they are, the target's actions may be predictable and even controllable—one of the most important facets to any ambush. For instance, if an ambusher knows that firing into the right flank will cause the target to automatically move to the left flank, that may suggest the advisability of setting up a demolitions killing zone along the left flank.

The good ambusher is concerned with getting to the ambush site, how long it will take to do so, and how long it is likely the ambushers will have to remain at the location, undetected.

The successful ambusher delineates the kill team, the search team, flanking security, protection team, and so forth. He or she carefully selects the types of equipment, arms, and explosives that will be necessary. In night ambushes, for instance, more tracer ammo should be loaded out. When an enemy reaction force or a police patrol is very near and immediate destruction of the target is required, the use of suppressed weapons may be necessary. Suppressed weapons give virtually no audible warnings to the target, and that can gain a precious few more seconds in which ambushers are the only ones firing, a few more seconds to kill targets and thereby reduce the volume of fire that will be returned.

The good ambusher has determined what types of standoff weapons and demolitions are called for, and what the ranges are likely to be in the killing zone.

Effective ambushes seldom "happen." They are planned to the minutest detail. Planning backward from the moment the ambush is sprung is useful in setting up the plan of attack. Planning forward from the conclusion of the hit to the time of arrival at base or safe area/house is the most efficient way of setting up the withdrawal plan.

The importance of planning in the ambush, and in all military operations, was capsulized by the Chinese military sage Sun-tzu when he wrote more than 2,000 years ago that "a victorious army first wins and then seeks battle; a defeated army first battles and then seeks victory."

Preparation for an ambush involves three elements: *planning considerations, intelligence,* and *selection of appropriate areas and sites.*

Planning Considerations

Planning considerations is a catch-all term that military and police use as a convenient "roof" for four interrelated issues: *mission, probable size, terrain,* and *timing.*

Mission

The first of the four is the mission. This may be a single ambush against a column or a series of ambushes against one or more routes of communication. In terrorist actions, it often involves knocking out a single car or two or three vehicles.

The mission really does affect the planning of an ambush. Take, for instance, the situation where the mission involves a harassing ambush of opportunity against the first vehicle convoy passing a point along a road. The plan is to halt the vehicles using electrically detonated demolitions, inflict casualties, and damage vehicles by using automatic weapons fire and electrically detonated claymore antipersonnel weapons. For this you need enough men to haul the explosives and claymores to the site and put them in place. You would probably keep a couple of men armed with automatic weapons and have the rest return to their base—if that could be done without being observed. In carrying out the ambush, you detonate the explosives, set off the claymores, and deliver a high volume of fire to the target. Before reinforcements can be brought up or the pursuit organized, the three-man ambush team withdraws.

Now look at a different mission: a deliberate ambush against a vehicle convoy to destroy all vehicles and kill or capture all personnel. Now you need personnel for a command element, assault element, and security element, as well as numerous specialized subelements. For instance, you are sure to need a supporting element just to provide the heavy automatic-weapons fire.

Or what about this mission: the patrol is operating in enemy territory, and a convoy must be ambushed to obtain supplies? In addition to your command, assault, and security elements, you will need a carrying detail to haul away the booty.

In an executive ambush, the mission will determine the size of the force, the weapons, and the plan. A planned kidnapping is different from a straight-out assassination.

Probable Size

The second consideration for planning an ambush is the probable size, strength, and composition of the ambush target; the formations likely to be used; and the reinforcement capability. A military formation on the move, a head of state in a security convoy, and a company executive unaware that anyone has even targeted him are vastly different targets that require different approaches.

Terrain

The third consideration—one that relates closely to another issue of area and site selection—is choosing terrain along the route that is favorable for an ambush. Particularly critical is the availability of unobserved routes of approach and withdrawal.

Timing

The fourth consideration is timing. This is a multifaceted consideration. Ambushes conducted during periods of low visibility, for instance, offer a wider choice of positions and better opportunities to surprise and confuse the enemy than do daylight ambushes. However control and movement to, and during, the night ambush is more difficult. Night ambushes are more suitable when the mission can be accomplished during, or immediately following, the initial burst of fire. Night ambushes generally require a larger number of automatic weapons, and they should be placed at close range.

In counterinsurgency operations, night ambushes are effective in hindering an enemy's use of routes of communication by night. This is important because friendly air cover can attack those same rebel routes during the day. The day-night, one-two punch effectively denies insurgents the ability to use the routes.

Daylight ambushes make command and control easier and permit offensive action for a longer period of time. A day ambush provides the opportunity to use the more effective aimed fire of weapons such as rocket launchers and recoilless rifles.

Terrorist attacks in an executive ambush are almost always daylight ambushes—in fact, they are normally ambushes conducted when the executive is on his way to work in the morning.

Intelligence

Intelligence is a key part of the preparation for an

ambush. There are some—including anyone from an intelligence section—who will swear that this is *the only* key part of preparing an ambush.

Inflated egos aside, they have a valid point. The ambush commander ideally will have maximum available intelligence on the target, opposing forces able to intervene, the population in the vicinity of the target area, and the terrain to be traversed en route to and on the way back from the ambush site.

An intensive—but not too noticeable (more about this later)—intelligence effort precedes the development of any plan. Ambushing forces reconnoiter the route and, if possible, the ambush site. Surveillance in the area is continuous up to the time of the attack. However, the ambush commander must exercise extreme caution, denying the target any indication that an attack is impending. Guerrillas often will be unable to determine in advance the exact composition, strength, and time of movement of convoys; their intelligence efforts will often be directed toward determining the convoy pattern of the target. Using that information, guerrilla commanders are able to decide on types of convoys to be attacked by ambush.

Terrorist attacks—depending on the target—may be able to predict almost every move of, and understand almost every facet about, their target. The intelligence-gathering process is a tip-off to targets, if detected. Careful observation, looking for the telltale signs of intelligence gathering, is one of the most effective ways of avoiding an ambush.

Selection of Appropriate Areas or Sites

When selecting appropriate ambush areas and sites, "A" stands for ambush. "A" also stands for anywhere! And any area where decisive surprise can be achieved is a good place. Clearly, some areas are better than others.

There are key characteristics to look for when choosing an ambush area—a consideration that any potential target should know and understand every bit as thoroughly as an ambusher:

- The target should be channeled.
- There should be good fields of fire.
- Good cover and concealment are needed for the ambush force.
- Where possible, there should be natural obstacles to prevent the target from reorganizing or fleeing from the site.
- Concealed approach and withdrawal routes from the site should be available to the ambush party.

Once the area for ambush operations has been determined, the actual ambush sites are selected. Favorable terrain is everything in an ambush . . . well, almost. Limitations that exist, such as deficiencies in the firepower available to the ambusher and lack of resupply during actions, may govern the choice of an ambush site.

Sun-tzu, in his treatise, warned that "when an army is traveling, if there is hilly territory with many streams and ponds or depressions overgrown with reeds, or wild forests with a luxuriant growth of plants or trees, it is imperative to search them carefully and thoroughly. For these afford stations for bushwhackers and spoilers." The world over, dense foliage and hilly country are known for their ability to hide bushwhackers and bandits.

The most successful ambushes are sited at locations that are designed to appear to be unlikely ambush sites, but which in reality give the ambusher a decided advantage while invariably placing the ambushed unit in an unfavorable position.

The "ideal" ambush site restricts the target on all sides, confining him to an area, a killing zone, where he can be quickly and completely destroyed. Unfortunately, it is seldom possible to prepare an ideal ambush site and completely restrict the target's movements. Natural restrictions or obstacles such as cliffs, streams, embankments, and steep grades that force vehicles to slow down should be used wherever possible. Artificial restrictions such as barbed wire, mines, and cratered roads are used not only to confine the target to the desired killing zone but also to inflict casualties. In Sun-tzu's time, the ambusher could take advantage only of natural obstacles such as defiles, swamps, and cliffs that would restrict the target's maneuver against the ambush force. But in modern days, even when natural obstacles do not exist, mines and demolition materials can be used to canalize the enemy.

The ambush site should have firing positions that offer both concealment and favorable fields of fire. Whenever possible, firing should be done through a screen of foliage. The terrain at the site should serve to funnel the enemy into the killing zone. The entire killing zone should be covered by fire to avoid dead space that would allow the enemy to regroup or organize resistance to the ambush.

As part of the planning, determine what form of transportation the enemy will be using. In desert terrain it is unlikely they will be moving far from a camp afoot. More likely, they will be traveling in vehicles. Know what type of vehicles they will be in, what it would take to stop the vehicles, if there are any means of rapid communications to supporting forces to lend them assistance, and how long it will take before those supporting forces arrive and provide assistance. In jungle terrain it is unlikely that the enemy will be mounted in tanks or trucks; they are likely to be on foot. But again, it is important to understand how long it will be before support will be available to the ambushed party—and whether that help will be other troops, air power, or artillery.

It is also important to know whether the enemy has effective counterambush tactics and whether the target can be expected to follow any counterambush SOPs. Using knowledge of enemy SOPs can increase the options available. For instance, an ambush commander who knows that fire into his right flank will cause the enemy to move to the left flank can set up a demolitions ambush killing zone on his left flank—and fire into the right flank.

Because security elements are placed on roads and trails leading to the ambush site to warn the assault element of the enemy approach, and those security elements will assist in covering the retirement of the assault element from the ambush site, terrain will be an important planning consideration. The proximity of the security elements to the assault elements will be dictated by terrain, and in many cases it may be necessary to organize secondary ambushes and roadblocks to intercept and delay enemy reinforcements—an issue that will require further careful consideration of terrain.

U.S. experts in Vietnam responsible for setting up ambushes against guerrillas recommended that:

> *. . . numerous night ambushes should be laid along railroads, roads, trails, and waterways which the VC must use to approach hamlets and villages. These likely approaches can be deduced if required intelligence is not known. Sites for ambushes can be found in remote areas by a close study of those locations where the VC contact the population while they are working in the fields. These ambushes should be set before dawn and prior to the arrival of the workers in the field. Because the VC leaves his safe areas to enter the populated areas, ambushes should also be laid along roads and trails and approximately 15–20 kilometers out from the perimeter of the populated areas.*

According to U.S. doctrine in the Vietnam era:

> *Ambushes are most effective when the site selected confines the VC to an area where he can be destroyed. Natural obstacles are numer-*

> *ous in Vietnam for ambush positions, such as cliffs, streams, embankments, and narrow trails and roads with canals on either side. An indirect approach should be used to enter the ambush site, otherwise the VC will detect the friendly movement and employ an ambush against GVN forces. At times use of a circuitous route may require three or four days to reach the ambush site. A patrol may be forced to occupy an ambush site well ahead of the arrival of the target. Patience is essential if secrecy is to be maintained. Therefore units must be prepared to remain in ambush areas for a minimum of a week and often as long as a month.*

A lot of fine words—which all too often didn't work in Vietnam.

Doctrine of other allies called for setting ambushes at:

- Known enemy routes, either in rear or forward areas
- Supply, arms, and ammunition and water replenishment points, as well as administrative locations
- Along probable lines of enemy withdrawal from an area
- At the interface of vegetation changes, e.g., where forest and grasslands meet

All were good ideas, but by themselves they could not and did not eliminate the VC and the North Vietnamese Army from the battlefield.

CARRYING OUT THE AMBUSH

Moving to the ambush site is a key phase of the operation. The force moves over a preselected route or routes. In large ambushes, one or more mission support sites are usually necessary along the route to the ambush site.

Up-to-the-minute intelligence is provided by reconnaissance elements, and final coordination for the ambush is made at the mission support site.

When approaching the ambush site, it is always best to come into the site from behind. The killing zone should be left completely undisturbed so that the target won't be aware that they are entering a dangerous area.

This is critical since studies show that the second greatest cause of broken ambushes is failing to enter the site from behind. All too often, when the ambushers pass through a killing zone instead of entering it from behind, they leave signs that cause the target to detect the ambush before it is sprung.

Planning should include enough time, and more than enough time, to allow the ambush party to traverse the terrain and distance and still be far ahead of the prey. Since the ambush party will want to be in harmony with nature, sufficient lead time is necessary to allow nature to return to its normal pattern and functioning after setup—the birds should feel free to chirp and the insects to set up their own cacophonous serenade. Where possible, the ambush team should arrive on site a matter of hours before the planned attack time. Four to six hours is the maximum amount of time ambushers should be expected to remain hidden. After that, even the best troops are less alert and circulation problems increase, diminishing the physical abilities of the ambush team.

Near the ambush site, troops are moved to an assembly area, and security elements take up their positions. Silence and immobility are scrupulously observed by the attackers.

This assembly area is formally known as a "lay up position," or LUP. The LUP, which should be close to the ambush site, has several uses. For ambushes of long duration or during hours when the ambush might be sprung unintentionally, the ambush party can fall back on the LUP. The LUP is also useful for holding anyone who accidentally stumbles upon the ambush and who is not to be eliminated. A properly chosen LUP will be one that is far enough from the ambush site that it will not be overrun if the opposition attacks/counterattacks the ambush. Although the distance will depend on terrain, cover, and a myriad of other factors, the LUP is often sited about 500 yards from the actual ambush site.

The ambush commander, the senior officer, or other person in charge of the ambush—sometimes called the patrol leader—should leave the ambush party at the LUP while he reconnoiters the ambush site itself. The ambush commander and the point man, who scouts a clear path for the ambush team, conduct the actual reconnaissance of the ambush site to choose good concealment for the fire teams. (Cover is not necessarily a consideration, only concealment.)

The ambush party should have a good view of the killing zone without being seen themselves. When preparing for an ambush during the nighttime, it is important that members of the attack party who may be well concealed in the darkness and low light are not placed where they will be exposed when day comes.

The ambush commander makes certain there is no cover in the killing zone. He makes certain there are no rocks, depressions, or heavy growth that will give the target an even chance.

Under no circumstances should the site be set up in such a way that an ambush *must* be sprung. An ambush should never be set where the ambush commander cannot allow the target force to pass by in the event that circumstances change (e.g., the approach of a larger force than anticipated) and the ambush commander decides the attack should not be carried out.

The ambush party is moved from the LUP with plenty of time to spare. Again, it is always best to enter the site from the rear so that the killing zone remains undisturbed and the enemy gets no clue that he is entering a dangerous spot.

Flank personnel should be in their positions before the

remainder of the kill team is put in place. From the first moments, it is important that the flanks be covered and flankers have a good view to see what is approaching—and doubly critical that the target not be in the killing zone before the ambush party is ready to initiate combat. Putting the flankers out early eliminates such problems.

The rest of the ambush team is positioned once the flanks are secured. Considerations that go into the placement of the attack team include:

- Ease of communication
- Control of the entire group
- Placement of the guide and radioman next to the ambush commander
- Knowledge of where every member of the attack team is located so that no member is in front of another's weapon or in his sector of fire

The weapons should be positioned where the fire sectors are to be. Once the target arrives on site, it is too late to make changes.

All automatic weapons, particularly the light machine guns, should have their left and right arcs of fire fixed to prevent anyone from shooting other members of the ambush team—particularly flankers—by accident. Sticks placed in the ground to designate the maximum permissible arcs of fire are effective.

By this point, issues such as sleeping patterns, wake-up signals, and communications signals that will be used should be established. Creature comforts such as mosquito repellent should be prepared prior to settling in to an "immobile" state.

Noise discipline is important when setting up the ambush, but it is important to remember that if any noise is to be made, it should be made when the ambushers first move in and set up the site. That's the time to clear out the local noisemakers—the sticks and gravel and such. The extent of clearing should not create a "comfort zone" that the target can recognize on approach, however.

Remember, keeping movement to a minimum at all times will reduce noise and make it more difficult for someone to see the ambushers. Movement attracts the eye!

Discipline is essential in any ambush. Sleeping, talking, eating, smoking, or chewing tobacco is dangerous, even foolish, at an ambush site. Where the ambush team has to be in place for an extended period, a sleeping plan can be decided upon, and those relieved from duty can sleep at another location, possibly the LUP.

When claymores and other explosives are being used—either for security or in or along the edges of the killing zone—there are important rules to follow. While one person places the mines, another must provide security. The flanks from which the target is most likely to approach should be mined one at a time. The rear security claymore is the last to be put in place.

In some cases claymores may be used to increase the size of the killing zone. In that case, it may be necessary to rewire the system, set delays, or otherwise customize the system. This should be done when initially setting up the charges.

After the mines are emplaced, signal (tug) lines may be set up between the members of the ambush team and the group goes into an immobile state.

The flankers' job is to signal if someone or something is approaching. Flankers positions are often set several yards from the main body of the attack team. Sometimes it is good to have flankers sit quietly while the hit is carried out, watching for any flanking maneuver. The flanker looks out for reinforcements and determines if the first element is a point element or part of the main body of the target. Having a flanker off to the side of the ambush site also has the effect of increasing the size of the killing zone in many cases.

Machine gunners are often positioned in the center of

the attack team so that they can cover the entire killing zone with fire. Their ammunition should be placed in a box, rather than on the ground, to help control noise and ensure reliability of the ammunition. In some cases the machine gunners will, when ordered to open fire, stitch a burst of all-tracer from one edge of the killing zone to the other as a final means of designating the size of the target.

The ambushers' radioman, if there is one, is usually the rear security. When illumination rounds are to be used, the radioman puts them up. The radioman should keep the set turned off or use only headphones except when it is needed. Incoming transmissions, if heard by the target, are a clear giveaway of the ambush. The guide is stationed next to the ambush commander.

When someone approaches, the first flanker to see the intruder passes the word, perhaps by pulling on a communication or tug line that is strung between the members of the ambush team. On feeling the tug, those in the main body of the ambush pass the information to the commander, again perhaps by tug line or squeeze. Nobody should react, except to pass the information. It is up to the ambush commander to initiate the action. The ambush commander should be waiting for additional information signifying that a larger force is approaching.

As the approaching target is detected in the killing zone, or at a predesignated time, the ambush commander decides whether or not to execute the ambush. Troops, particularly if the ambush is at short range, often tend to stare at the approaching target and fix it in their sight and gun sights. Experienced ambush commanders insist that, for reasons unknown to science today, people can and do sense the "presence" of others who are staring at them. For this reason, to prevent the possibility of tipping off the enemy through processes that are not understood fully, it is best for the ambush team to view the oncoming enemy indirectly, in peripheral vision. Certainly staring at the enemy doesn't help—and seems to have a negative effect on vision at night, particularly.

It is important to stress that it is *always* the responsibility of the ambush commander or his designee to decide whether to attack the target. In the case of a military convoy, the commander's decision to attack might depend on the size of the column, the extent of the guard and any security measures, and the estimated worth of the target in light of the mission.

The commander determines when to open fire. In close terrain, this may be, but is not usually, on sight. In open terrain, it may be on a predetermined signal or when the target has reached a predesignated point. The "when" is always a matter of judgment, but the most successful ambushes are those in which the target is well within the killing zone. Not only will a premature triggering of the ambush likely result in a failure of the attack, it may well jeopardize the safety of the ambush party. In ambushes other than a demolition ambush, if at least 90 percent of the target can't be gotten into the killing zone at one time, the ambush should not be sprung. When more than 10 percent of the enemy force is outside the killing zone, the enemy has too much freedom to flank and use fire and movement to defeat—and even destroy—the ambushers.

Depending on the setup and circumstances, the ambush commander can initiate action with a weapon or claymores. Nothing other than a casualty-producing weapon should ever be used to initiate the ambush. Pyrotechnics, whistles, etc., only give the target more time in which to react. Once the ambush is initiated, the other members of the attack team bring up their weapons from the position where they were being held at the ready, remove the safety, and fire. The safety should not have been removed at any earlier time.

(Some will disagree with this and insist that all safeties should have been put off when the ambushers went into an immobile state. Unfortunately, the chances that someone will fire off a round accidentally are generally too great, except in the most disciplined of troops, to make this approach feasible. Another approach calls for safeties to be removed at some point between the initial warning signal from the flankers and the commander's signal to open fire. This is also a good time to remove safeties—except when the target is a slowly moving foot patrol or unit on a march. Then the massed sound of safeties being removed might be audible to well-trained troops, scouts, or point men. It might also frighten bird and wildlife and set off a cacophony that will alert the ambush target to potential danger. However, when attacking mechanized targets or those in them, the sound of the safety or many safeties being clicked off is unlikely to have an effect.)

Sectors of fire into the ambushed enemy should have been predesignated to ensure that the entire force is smothered by fire. Enfilade fire is desirable. If the ambush target is a vehicular column, initial fire is delivered at the front or the rear vehicle, whichever is at the weakest point of the ambush site. If both front and rear are equally strong, the first fire should be directed at the trailing vehicle.

If the decision is made to execute the ambush, any advance guards are often allowed to pass through the main position. But when the head of the main column, or the high-value target, reaches a predetermined point, it is halted by fire, demolitions, or obstacles—including felling trees into the path. At this point, the entire assault element opens fire. Designated details engage the advance and rear guards to prevent reinforcement of the main column.

Attacks are made at close range; this compensates for poor marksmanship and gains maximum effect. Automatic weapons are often used to cover the entire target in depth. Shotguns and grenades may be used, as well as area weapons, including flamethrowers. In any event, ambush teams should be well rehearsed in using their weapons so that they are able to maintain a maximum, sustained volume of fire.

In night ambushes, illumination may be useful since it helps focus fire on the enemy. Targets should be illuminated with flares from mortars or pop flares and should be placed low and behind the killing zone so that targets are silhouetted. In that location, the flares are unlikely to illuminate the ambushers. The ranges will have to be determined by the type of flare used, and the flare sequence should be carefully timed. For instance, one flare should be bursting while the second is in mid-burn, and the third is about to burn out or land behind the target.

While the flankers look for reinforcement or flanking maneuvers, the riflemen hit any point targets on their zone—where possible, one round, one hit. Machine gunners cover the entire killing zone; grenadiers place rounds along the trail or route. The radioman provides illumination and may call for extraction or fire support when instructed by the ambush commander.

The volume of fire is rapid and directed at all enemy personnel, all exits from vehicles, and automatic weapons. Antitank grenades, rocket launchers, and recoilless rifles are used against armored vehicles. Machine guns lay bands of fixed fire across possible escape routes. Mortar shells, as well as hand and rifle grenades, are fired into the killing zone. The last two or three rounds in all rifles and automatic weapons should be tracer so that the shooter knows when he is about to run out of ammunition and can prepare to change magazines or insert a new clip.

The ambush commander provides fire support by shooting tracers at targets. He also gives the order to lift, shift, or cease fire.

In some cases, before everyone has been killed at an ambush site, the plan and the commander may call for an assault on an ambush target. In that event the attack is launched under covering fire on a prearranged signal.

The ambush commander may also terminate the ambush for any of three very good reasons:

1. The enemy has been annihilated.
2. The enemy has been routed and is retreating.
3. The ambush is unsuccessful.

If the ambush commander believes everyone is dead, he commands "cease-fire" and calls for "security out." The rear security stays where it is; the flankers stay in position. In an on-line ambush, the machine gunner or gunners move across to the far side of the killing zone. Security should cover the compass—it should be 360 degrees or as close as possible to it. On the order of the ambush commander, the assault squad may walk through the ambush site a short distance to the other side of the killing zone, searching for wounded or escaping members of the targeted group. Flankers have a 180-degree field of responsibility during this phase.

Clearly, there may be reasons for people to shoot during this walk-through. Just as clearly, any gunfire at that point is a cause for alarm and can set off an uncontrolled response—with people firing in all directions. For that reason, anyone shooting while in the perimeter or during the walk-through of the killing zone needs to let everyone else know what he plans on doing. "Going hot" is the generally recognized signal that someone needs to open fire.

Any search team is sent out after the walk-through. Usually composed of two people, one ensures the kill while the other searches. Both parts of the team need to be constantly aware of explosives. While watching for hidden grenades, booby traps, or people playing possum, the searcher cuts away clothing and removes items.

Generally, a head count should be conducted before withdrawal, and ammunition should be counted and, where necessary, redistributed.

WITHDRAWAL FROM THE AMBUSH SITE

When the commander desires to terminate the action because the mission has been accomplished or superior enemy reinforcements are arriving, he gives the order to withdraw to an extraction point or rally point. The first to withdraw is the assault element. When living enemy troops remain in or near the killing zone, the ambush team pulls back in pairs, or "buddies." The buddies leapfrog out, one attacker covering the other as they withdraw along preplanned routes to the rally point and LUP. The automatic weapons remain in place through the initial part of the withdrawal and keep up sustained fire area-wide and at targets of opportunity during the pull-back. The security element follows the assault element from the ambush site.

Tear gas and chemical smoke may be used to cover a withdrawal when wind conditions and circumstances—such as the enemy breathing down your neck—permit or require.

The ambush team may not be withdrawing from the site under fire, however. At times the ambush will have been so successful that there is no one who can effectively resist in the killing zone—they are either dead or badly wounded—and search teams are sent out. Once the search team has completed its task, the flankers may crank off the claymores. If not, then the claymores are removed as they were emplaced—one man doing the work while the other provides security. If time-delay demolitions are planned to catch possible pursuers, this is when they should be put in place.

This is the time to move the ambush team back to the LUP. When withdrawing, the idea is to achieve maximum deception of the enemy and facilitate further action by the force. The various elements withdraw, in order, over predetermined routes through a series of rallying points. Should the enemy mount a close pursuit of the assault force, the security element assists by fire and movement, distracting the enemy and forcing him to slow down. Under normal circumstances the routes of withdrawal are not the same as those used to reach the ambush site originally.

Elements that are closely pursued do not attempt to reach the initial rallying point, but on their own initiative lead the enemy away from the remainder of the force. Where there is difficult terrain, they use that to attempt to lose the enemy "tail." When the situation permits, an attempt can be made to reestablish contact with the remainder of the ambush force at other rallying points or to continue to the base area as a separate group.

When necessary, the ambush force, or elements of it, separates into small groups or even individuals to evade close pursuit by the enemy. Frequently, even if there is no close pursuit, the ambush force disperses into smaller units, withdraws in different directions, and reassembles at a later time and predesignated place. They may conduct other operations during the withdrawal, such as an ambush of the pursuing enemy force.

COUNTERAMBUSH

There is another viewpoint to consider the side of the target.

Once an ambush is detected—sometimes because of poor noise discipline but all too often just because the ambush is sprung—the problem becomes a matter of moving against the enemy in a manner that will force him to abandon his position. If possible, the counterattack should move the ambushers into an area where maneuver and supporting fires can quickly and easily defeat him.

If ambushed, all troops (and civilians who are at risk) must be conditioned to react immediately and violently, without orders, to overcome the initial advantage held by the enemy. The immediate-action drills are one example of this reflex type action. Because an ambush inflicts its casualties almost immediately and no attempt is made to prolong the engagement, an immediate reaction to build and retain fire superiority is the best initial defense against the ambush. The use of automatic weapons, fragmentation, smoke and white phosphorous hand grenades, and small arms fire will tend to win the initiative from ambushers. Flame weapons, canister shot from large-caliber weapons, and vehicular-mounted rockets are devices that can assist in military ambushes. In executive ambushes, the use of the vehicle as a weapon may help break up the attack. A prompt, decisive initial reaction will materially reduce casualties and the ultimate effectiveness of the ambush, but, in itself, it is not enough.

The weak spots of the guerrilla organization must be sought out and attacked during the response to the ambush. As previously mentioned, command and control in the ambush is essential. The ambush commander will usually position himself where he can best control the ambush, at its trigger, so to speak, the point where the head of the column must be stopped. Heavy fire directed at the point might cause him to become a casualty, thereby disrupting command and control. Some degree of caution must be exercised in this regard. A battle-wise enemy will often ambush, wait for a reaction to develop against the ambush party, and then strike from a different direction with the primary ambush force. Hence, only sufficient firepower should be employed to gain and maintain fire superiority; the remainder should be reserved for the primary ambush, or, if none develops, the firepower should be employed to support elements that maneuver against the ambush position.

Since the success of the ambush is predicated on simple signals used by the ambushers for command and control, duplication of signals known to have been used previously in ambushes might confuse attacking forces, causing them to cease fire prematurely, withdraw, or assault. This reuse should not be neglected, if known.

When ambushed, individual and unit "reflex action" to initially overcome the ambush must be followed by violent and aggressive counteraction. Since the ambusher fears denial of escape routes or a flanking attack by an organized force, every effort must be made to rapidly begin an envelopment of the flank of the ambush party. Preferably, this should be accomplished by an element of the unit not engaged in the ambush. The flanking move can expect resistance but usually not determined resistance.

Once the flanking threat is known, the ambushing unit will normally attempt to break contact. For that reason, such flanking action should be made in coordination with a frontal assault by those troops caught in the killing zone of the ambush—but only after fire superiority is achieved.

Once the ambushers attempt to break contact, pursuit operations must be launched immediately; relentless pursuit is a must, and contact must be maintained. That is the advice, and it is good advice in some ways, still given by armchair theorists. It was the advice given in Vietnam, and trying to carry it out resulted in needless casualties. In real-world operations, following the advice of such armchair warriors can be dangerous. The phrase should be "relentless, intelligent pursuit."

It is all too easy for the pursued to lay some kind of a hasty ambush for the pursuers. If the ambushers are running across open fields, where they can be seen and any new ambush attempt can be spotted, relentless pursuit might make sense. But in heavy growth, such as a triple canopy jungle, relentless pursuit afoot is foolish, not to mention potentially fatal.

A far better method is to put just enough pressure on the fleeing ambushers to keep them moving, while setting blocking forces up ahead of them. The use of helicopters to position forces ahead of the withdrawing ambushers will normally eliminate delaying ambushes.

Supporting long-range weapons can bring their firepower to bear on the attackers, as well. These methods, too, are other forms of "relentless intelligent pursuit."

Air support designed to kill is a must during the day. During the hours of darkness, air support can illuminate the area of pursuit.

The underlying rule is that, to the extent possible, all opportunities for escape must be denied to ambushing forces. That makes sense. And immediate-action drills, as well as artillery and air support, can help in preparing troops to react positively against ambushers.

Since any military unit caught in an ambush must react without the slightest hesitation in order to avoid annihilation, counterambush techniques are conducted in training as immediate-action drills. They are practiced until every member of a unit will respond almost instinctively when first fired upon. The military techniques for counterambush apply to civilians who have a security team and, in some cases, even a single bodyguard.

In the military world, as mentioned above, the two generally recognized techniques for effectively countering an ambush are as follows:

1. The element that receives the initial burst of fire takes cover and immediately returns a maximum volume of fire—whether that's from a single handgun or 100 barrels. As this element attempts to gain fire superiority, the elements that have escaped the initial burst of fire immediately begin pre-drilled maneuvers against the flanks and rear of the ambush without further orders.

2. The element that receives the first fire returns that fire and immediately assaults the ambushers' positions. This technique requires extensive training and unusual alertness. However, this action tends to astound and confuse the enemy and, in most cases, will cause him to panic at the apparently mad and reckless action, thereby achieving success with fewer casualties. Speculatively,

such a course of action will seriously hinder the attacker's withdrawal and may force him to remain in the area while reinforcements arrive.

These techniques are fine for military units or even civilians with bodyguards, but they will not be effective for the unarmed executive who is by himself. He can only spit and throw stones—if he is still alive after the first burst of fire. For him, or in some cases her, there is a third course of action.

3. Get out of the killing zone any way possible. Run, crawl, or drive through barricades. When you're outgunned, no amount of personal courage can compensate for the difference in firepower.

PREVENTION OF AMBUSHES

The ambush is not the ultimate military or terrorist tactic; it is not a methodology against which there is no defense. Even though terrain, weather, state of training, and the equipment of the enemy may favor this tactic, there are proven measures and countermeasures that can be taken that will materially reduce the incidence of effective ambushes. Suggested preventive measures:

1. Since ambushes are predicated upon accurate intelligence of routes, timetables, escorts, and locations of communications equipment and leaders, counterintelligence activities should be intensified. Denial of the requisite information will reduce the number of effective ambushes. Many deceptive measures can be taken to deceive the would-be ambusher as to the true nature of the movement. Leaving departure areas with a larger-than-required escort that drops off en route will tend to deceive the ambusher's intelligence efforts. Variation of formation to change the locations of communications, leaders and automatic weapons within the unit will further confuse intelligence efforts. These variations may simply be formation changes

for different movements, or they may be changes during the conduct of one particular movement.

2. Civilian cooperation, or lack thereof, is a key to preventing some ambushes. Many irregular forces, like the VC, normally insisted on maintaining routine civilian activity in areas where ambushes were laid. Therefore, the presence of "normal" activity should not be construed as an absolute indication of safety. Civilians indigenous to the areas are often aware of ambushes that are set. Careful interrogation of some of them may reveal the presence of the ambush, but that shouldn't be counted on. When it is suspected that local inhabitants have knowledge of an ambush, they may reveal the presence of the enemy by their reluctance to enter the known ambush site with the lead scouts of the friendly unit.

3. The use of helicopters and other type of army aircraft to conduct reconnaissance, as well as to reinforce and position flank security units, will tend to reduce ambushes.

4. Since an ambush, ideally conducted, is not discovered until it is sprung, the most effective preventive measure is the employment of adequate security. Alert scouts will detect ambush sites by searching out suspected areas, observing signals that herald their arrival, or by drawing fire. In this regard, when moving along trails, movement must be made in complete silence. Ambushers often occupy ambush sites for prolonged periods, and they tend to become careless about noise discipline. Under such conditions, lead scouts can often hear the ambush party and circumvent the ambush site if the main body conducts its move in silence. Units can also employ *controlled* reconnaissance by fire—including preplanned artillery concentrations—and move by bounds, covering the advance by mutually supporting automatic weapons. Armored vehicles can carry security detachments and carry out mounted or dismounted flanking action. Any action that will cause the ambushers to either spring the ambush prematurely or reveal its

presence will greatly reduce the effectiveness and number of ambushes. In convoy situations, reconnaissance units should travel ahead of the main body, attempting to prematurely trigger potential ambushes.

5. Because the ambush requires a cohesive organization with close coordination designed to silence certain key defenses early on, random changes in the organization of the column, the location of supporting weapons, the communications equipment, etc., will require similar changes in the organization of the ambush force. If schedules are varied and changes in patterns and formations are made at roadside halts, the ambush force may not be able to complete its reorganization in time to engage the target properly. Either the ambush commander will permit the targeted unit to proceed without attack, or the changes will render planning for the ambush ineffective.

6. Units moving to relieve an outpost unit that is reported to be under attack must be especially vigilant to avoid ambush since this is a classic tactic for setting up an ambush. Ideally, relief forces should move by helicopter to the outpost; forces moving overland should seek to locate, fix, and destroy the attack party while dispatching a suitable relief to the post itself. Priority should be given to destruction of the ambush/attack party. Overland movement should have air cover and employ all the applicable techniques to detect any ambush or reduce its effectiveness and pursue the ambush party relentlessly until it is destroyed.

7. Units returning to "home stations" must be especially alert for ambushes. Return movements, by either foot or motor, must use alternate routes. The easiest or most logical alternate return route should be avoided if possible. Ambushing units have sometimes been in place for an attack but nonetheless permitted the outward movement of opposing forces only to ambush them as they returned. This course of action capitalizes on the natural tendency of troops to lower their guard on return

trips, especially when traveling over a route that has been used safely a short time before.

8. Although there are other methods of preventing ambush, the fundamental principle is *security*. Security in this sense includes all measures, both active and passive, to provide for the safety of the command. Depending on circumstances, it may include such diverse elements as sniffer dogs, picket boats, vehicular-mounted flamethrowers, mounted patrols, supporting weapons, air support in the form of column cover, and an especially alert rifleman at the point. No available and appropriate measures should be omitted. Movement must go on. However, it must be movement that is properly designed in order to prevent effective ambushes.

• • •

While these points apply particularly to military operations, they are applicable in the civilian world as well. For instance, the most dangerous part of a businessman's or official's trip is the 200 yards in either direction from home, just as the last lap back to an outpost is the most important.

2 Special Situations

AERIAL RECON

Deficiencies in the conduct of aerial reconnaissance were not unknown in Vietnam. In fact, bad aerial recon work is worse than none at all when it comes to setting up anti-ambush operations because it can tip off the enemy that something is in the wind. Government documents note that "in one operation it was determined that the VC had left the area which they had occupied for some time just before the operation was launched. Although it cannot be proven, it was the advisor's assessment that extensive and unusual aerial reconnaissance prior to the operation had alerted the VC."

Deception must be employed when conducting aerial recon. This can be achieved by overflying other areas in addition to the objective area and by the use of aircraft normally used in the sector. Further, as many leaders as possible should be on the initial reconnaissance flight to preclude the need for additional overflights of the area.

In some cases where friendly troops are preparing an offensive operation, faulty aerial recon can prompt ambushes. Repeated reconnaissance of primary helicopter landing zones (LZs), particularly, can become a problem. If noted by the irregulars against whom the unit is operating, such recon missions can give them an advantage. The advance warning allows time to plan, prepare, and execute an ambush at the critical time and place, the LZ. As a key security measure, several alternate LZs should be selected in the preliminary phase of a heliborne operation. Selection of the primary LZ should be delayed until as late as possible. As a general rule, the primary LZ should not be selected more than eight hours prior to L-hour.

ALTERNATIVE STRATEGIES FOR CONVOYS

A major U.S. study of VC ambushes recommended that

when a convoy was ambushed, "the proper immediate reaction for vehicles caught in the killing zone is to drive through the ambush if at all possible. Personnel on vehicles that have been permitted to drive through the ambush zone and those not yet in it should dismount and attack the ambush position from the flanks. If forced to halt in the killing zone, under current doctrine troops dismount at once and immediately assault the ambush party."

The study warned that perhaps the tried and true, cut and dry policy of immediate attack was not the best answer. The new doctrine in later years offered alternatives to the expert soldiers—and perhaps to that extent confused the situation:

> *Experience to date in Vietnam . . . indicates that equipment and terrain considerations as described below offer alternate courses of action which, if followed, could reduce casualties and enhance the possibility of successful counterambush operations. For example, under certain conditions, it may not be desirable to detruck immediately and conduct a frontal assault against an enemy who has the advantage of surprise, fire superiority, and prepared positions.*
>
> *The initial opportunity for an alternate course of action occurs when protection has been provided for vehicular-mounted troops against the initial heavy volume of fire. Reaction could then proceed as follows: first, if vehicles are equipped with area-fire type weapons, these weapons, plus all rifle and automatic weapons available, should be used immediately to deny or reduce the effectiveness of VC fire superiority. Second, whether or not these area-type weapons have been provided, personnel should remain on the hard-*

ened vehicle, return the enemy fire with weapons available, and seek to survive until the first perceptible slackening of VC fire. This "lull" may be due to fire being masked by the enemy assault group, to a weapon reloading requirement, or to any one of a number of reasons. As soon as the lull occurs, however, the ambushed unit must take immediate action to seize the initiative from the VC.

While initially the tacticians thought that it was most important to get troops off armored or sandbagged vehicles—to spread them out—a second analysis led to a change in tactics.

> *When forced to halt in the killing zone of an ambush in a "hardened" vehicle, detrucking should NOT take place as currently prescribed. Instead, detrucking should occur:*
>
> *1. During the first lull in the initial intense VC fire (and under the cover of return fires from other personnel in the hardened vehicle) or*
>
> *2. Immediately following the delivery of a barrage from area-fire type weapons (which may preclude the VC from gaining and/or maintaining initial fire superiority).*

The new rule became stay put, fire back, and move for other cover and concealment only after the bad guys let up on their fire—either because they had to reload or needed to keep their heads down because of the hot lick of flamethrowers or the rain of machine-gun bullets.

> *The second opportunity for an alternate course of action occurs in the choice of tactics*

to be used by the troops in the killing zone after they have detrucked. Current Republic of Vietnam Armed Forces (RVNAF) doctrine prescribes an immediate frontal assault. Such tactics may be required under certain conditions, e.g., where flat, cleared terrain gives no opportunity for cover between the ambush force and the VC, or when VC ambush positions are immediately adjacent to the killing zone. Frequently, however, terrain, structures, or other features provide cover from which the time- and battle-tested techniques of a base of fire and maneuver and/or flanking tactics can be employed. In this case it appears more logical after dismounting to take cover, build a base of fire, and employ a maneuver element against ambushers.

The study showed that the lessons regarding the importance of taking cover when available, building up a base of fire on the ground, and employing maneuver elements against the ambushers had equal application to troops who have dismounted from a soft vehicle (one without armor protection) or to a foot column caught in an ambush. The preferred action in both instances was maneuver under the covering fire of individual weapons. Only if no cover is available should the immediate frontal assault against the ambush party be employed. But, in this regard, the doctrine writers still contended that the most effective counteraction to ambushes continued to be a flanking attack by any elements not in the killing zone and a relentless pursuit of the ambush party.

AMBUSHING WITH PATROL CRAFT

When conducting operations against boat traffic offshore, patrol craft operations are often very obvious. The sound and fury of high-speed patrol craft pelting up to interdict vessels and search them for contraband give those aboard time to hide or deep-six any incriminating materials and documents. Rebels and their sympathizers often conceal their activities while in proximity to a patrol boat. However, patrol craft can deceive the enemy into operating in the open, where they can be attacked. Deception can overcome the ability of the enemy to see and hear the approach of patrol craft.

There is a way to set up a waterborne ambush with patrol boats. A pair of patrol craft approaches the beach close together at night. One turns its engines off while the other breaks, at high speed, for open sea. Fishermen are deceived into thinking there was only one patrol craft and that it has now departed the area. The second craft conducts covert surveillance with night vision goggles, infrared detection equipment, and an occasional scan of radar on batteries. Any suspicious craft are reported to the patrol craft off the coast, which makes the contact without revealing the position of the other patrol craft. A related tactic is for a patrol craft to launch a Boston Whaler prior to entering an area of suspected enemy activity, then make a run through the area. After the patrol craft noisily departs the area, the Boston Whaler follows along the same track and reports suspicious activity.

The U.S. Navy and Vietnamese forces developed decent tactics in Vietnam. But that doesn't stop other rebels from using naval tactics in their operations. More recently, Sri Lankan forces have enjoyed similar success in ambushing rebel boats making night crossings of Sri Lanka's northern Jaffna Lagoon. The rebel naval force—the Sea Tigers, as the guerrillas are known—operate in wolf packs of high-speed fiberglass dinghies. The boats are powered by multiple outboard motors capable of reaching speeds of 30 knots, and they mount guns and rocket launchers. Each boat usually carries five or six guerrilla fighters and, en masse, they are formidable opponents. Even aircraft and other government patrol craft have a hard time operating against the Sea Tigers when they move as a large force.

ARMORED CIVILIAN VEHICLES

Most executive ambush situations take place in the car, in the morning on the way to work. There is nothing anyone can do about morning travel or about going to work. Mornings come; people have to go to work sometime. But there is something to be done about the vehicle itself. Thus, the armoring of the executive's travel car and any follow cars used by security personnel is highly desirable. Although armoring vehicles is expensive, it is not as expensive as losing a life.

The theory of vehicle armoring is simple. The car or other vehicle, such as four-wheel drive, should be able to absorb an initial burst of gunfire in an ambush and still remain maneuverable, the occupants uninjured. An armored vehicle should also provide some safety in a bomb attack. Still, there is no vehicle that can withstand a Lebanese Car Bomb attack if the ambushers build a big enough charge and detonate it close enough to the target vehicle. Few armored vehicles can take a direct hit from military-style armor-piercing shells or rockets.

Vehicle armor is generally divided into light and heavy armor. Just as there are threat level standards in the sale and use of body armor, or so-called bulletproof vests, there are standards for vehicle armor.

The armor used on a vehicle should be determined by the threat level—are potential ambushers likely to use a .357 Magnum or a 30.06? It makes a great deal of difference in the type of armor that will be needed. The level of armor needed can usually be determined by an examination of past attacks made by the group deemed most likely to stage an executive ambush. The rule is as follows:

always armor against the specific weapons most likely to be encountered.

Where the possibility of an assassination attempt is likely, or in cases where serious threats have been made against the life of an executive, heavy armor is called for. Heavy armor is designed to allow the occupants to button up and call for help. It is *very* expensive. It also puts great strains on the vehicle and the driver. The heavier the armor, the more strain it puts on the vehicle's suspension system. The car's maneuverability is seriously impaired by heavy armor—maneuvering one is literally like driving a tank. In hilly areas it may be difficult to get up speed on the ascent, which means that in any ambush set on an upgrade the vehicle will be in the killing zone longer. Because it is heavier, it will be harder to brake to a stop in an emergency.

Armoring is a specific art/science and cannot be covered in detail here. However, it is safe to say that generally a car selected for armoring should:

- Be similar to others on the road in the area where it will be used. It should not stand out or invite attention.
- Have four doors. This allows ease and speed of entrance, as well as comfort.
- Have sufficient room for four people to sit comfortably.
- Have a heavy-duty engine capable of meeting or exceeding the demands that will be placed on it. Engine size and gear ratio should allow for reasonable acceleration when it is fully occupied.

Modifications will almost certainly have to be made to the chassis and suspension system, particularly if the gas tank is to be protected by armoring. Where possible, a diesel engine is preferable to one that is gasoline-powered. Diesel fuel is not as explosive or prone to catch fire in an ambush and fry the occupants, although a diesel generally cannot accelerate as quickly as a gasoline engine.

Engines are marvelous examples of an engineer's skill. They will often run under seemingly impossible conditions. The armoring needed for the engine depends, to a great extent, on where the car will be used. In most urban ambushes, where attackers are not likely to pursue a vehicle, even a badly damaged engine will allow the vehicle to get out of the killing zone. In rural situations the engine compartment needs armoring because pursuit is more likely and the car has to travel farther to reach a safe or semisafe location. The one thing that will stop a vehicle faster than just about anything is a punctured radiator. In areas where it is likely the car would have to go many miles after breaking out of an ambush, an armored radiator is a necessity.

Heavy-duty and/or redundant power-steering pumps are needed on all armored vehicles. When a driver has to make evasive maneuvers, for whatever reason, standard pumps may not be able to perform. A loss of maneuverability can be fatal.

Communications should be assured so that help can be summoned immediately in case of ambush. A car phone is a must. In cases of high-threat levels, a redundant communications system, one using covert antennae, is needed. The communications system, of course, will be dependent on what is available in the local area. Cellular phones are operable in most areas of the world today, however.

A public-address system, for lack of a better term, is highly desirable. With a two-way system there is no need to open the door and breach the security of the vehicle in order to talk to someone. In fact, the driver doesn't have to be near a person to talk. The car can be kept at a distance for conversation. A good system will also allow the car's occupants to hear what the people outside are discussing among themselves.

A good mirroring system is a requirement for an armored vehicle, not only for driving safety but so that the driver can be aware of potential ambush situations developing to the side and rear.

All cars should have tool kits for roadside emergencies, but in the case of vehicles that may be involved in an ambush designed to kidnap the driver, the selection of tools in the trunk should allow a captive to free himself and escape if he is bundled into the back.

Finally, but most importantly, window armor should be substantial enough to withstand the same type of attack as the car body, in theory. In reality it is hard to get glazing that is totally effective against repeated strikes with high-powered weapons. Although it seems almost ludicrous, there have been cases where the body was armored and the window glass was the standard from-the-factory glazing. It is also important to keep in mind that window armor all too often distorts the view unless it is of the highest quality. That makes it pricey—but armoring any vehicle to any threat level is expensive. In some recent incidents, the armored glass was either down or had a space in it, and the perpetrators were able to place rounds through the openings and assassinate the occupants of the vehicle.

ARMORED ESCORT VEHICLES

The armored escort vehicles in a military convoy serve two important functions during an ambush. First, they provide immediate protection for personnel caught in the killing zone by driving into the ambush and engaging the attackers at point-blank range. Second, they provide direct fire support for a subsequent counterattack. Armored escort vehicles can be distributed throughout the line. When large convoys move on main roads, vehicles should be divided into blocks of five or six with an armored escort vehicle placed between each block. The armored escort vehicle may serve as a mobile command post for the convoy commander.

AVOIDING RIVERINE AMBUSHES

Both civilian vessels and police/military patrol craft—ranging from rubber rafts to diesel-powered boats—are at risk of ambush in the riverine environment. Significantly, ambushes along a river or waterway are among the least logistic-dependent of all ambushes. They can be staged with few if any rockets, small amounts of ammunition, and little or no explosives.

The term riverine ambush covers a great many types of attacks. An ambush of boats conducted by a patrol craft concealed along the edge of the waterway is a riverine ambush; so is an ambush carried out by ground forces stationed along the banks of a river.

The climate along most waterways is temperate, a factor that favors ambushing forces because they can remain immobile, in place, for lengthy periods of time without being unduly uncomfortable.

The terrain usually favors ambush, as a tactic. A river usually has many locations where there is decent cover and concealment, and the fact it is a waterway generally means that there is a clear field of fire. Because there is a natural channel that widens and narrows, often significantly, rivers have natural choke points that can be easily exploited.

A good waterway ambush is easiest when the cover afforded by the river environment is used well—in other words, when high banks, trees, bends in the river, islands, and rocks are carefully employed. Ambushes of riverine forces, whether patrol boats or civilian craft, most often occur near bends in the waterway. At bends in the river, control of the craft is more difficult, return fire is less easily massed, and the river channel general runs closer to one shore than the other. Islands and obstruct-

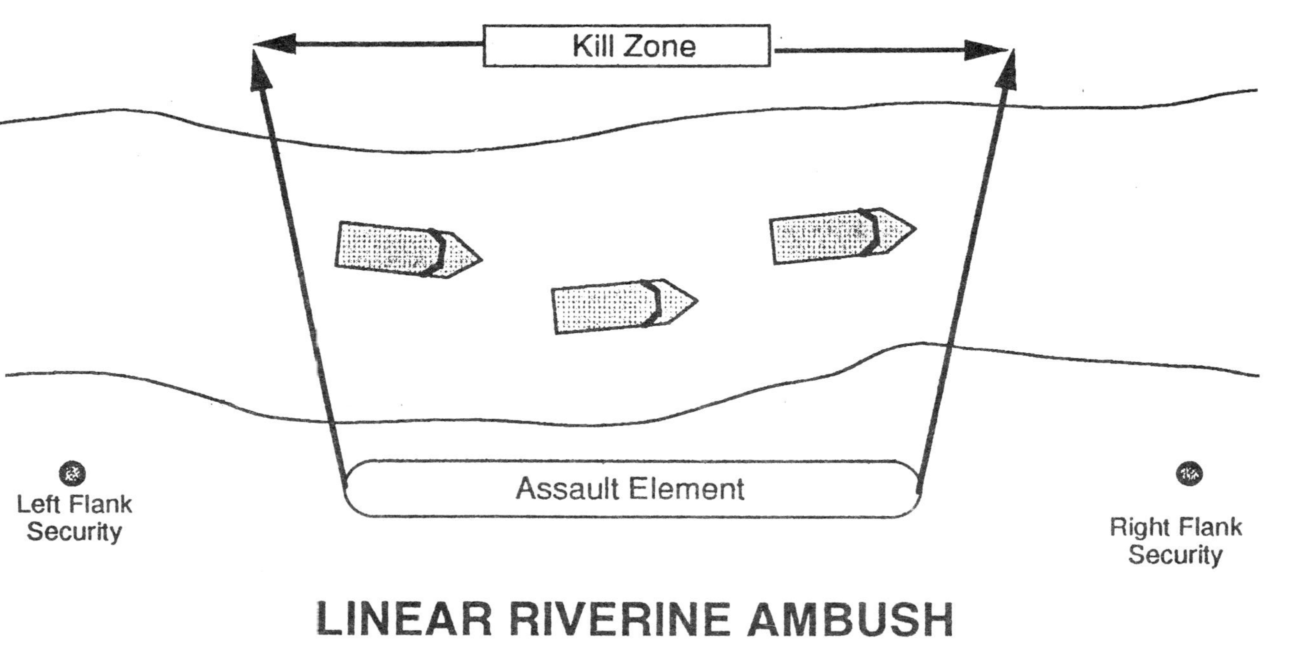
Kill Zone
Assault Element
Left Flank Security
Right Flank Security
LINEAR RIVERINE AMBUSH
Rear Security

ing rocks in the waterway channelize boats into waters where maneuvering is severely restricted.

Cover and the correct emplacement of personnel, weapons, and demolitions are the key to successful riverine ambushes, not the quantity or quality of the weapons. A key point to remember is that the smallest, most poorly armed group can mount successful river ambushes—for example, the fundamentalists in Egypt who so successfully attacked the Nile tourist traffic in 1992–93.

Riverine ambushes—which includes attacks on watercraft plying rivers, sloughs, and streams—can be extremely simple. Yet they are often complex. For instance, riverine ambushes are often multiple-pointed. In other words, a series of ambushes is mounted from the shore using rockets, command- or contact-detonated explosives, and automatic weapons. Typically, the boat or boat patrol moves into the first ambush, and if it is armed, it tries to shoot its way out. Whether armed or not, it will attempt to speed from the killing zone. In a multipointed ambush, as the guns and weapons aboard the boats are brought to bear against the first ambush position, a second ambush—either above or below the boats—is initiated. Now the fire of the defenders is divided. The boats now attempt to react to both of the ambushes, shifting fires or speeding away, when yet a third ambush is sprung, often in the direction the boat is moving.

In very narrow water if there is more than one boat or small craft in a convoy, and craft are sunk or sinking from the first two attacks—thus blocking retreat or advance out of the ambush—a third ambush between the original two can make the stretch of water a true killing zone since it bottles the craft up.

In any event, the multipointed ambush tends to keep defenders constantly shifting and off-balance, both psychologically and from a command and control standpoint. The order of the series of waterside ambushes can be in any sequence—they can even be strung out like

AMBUSH BARRIERS

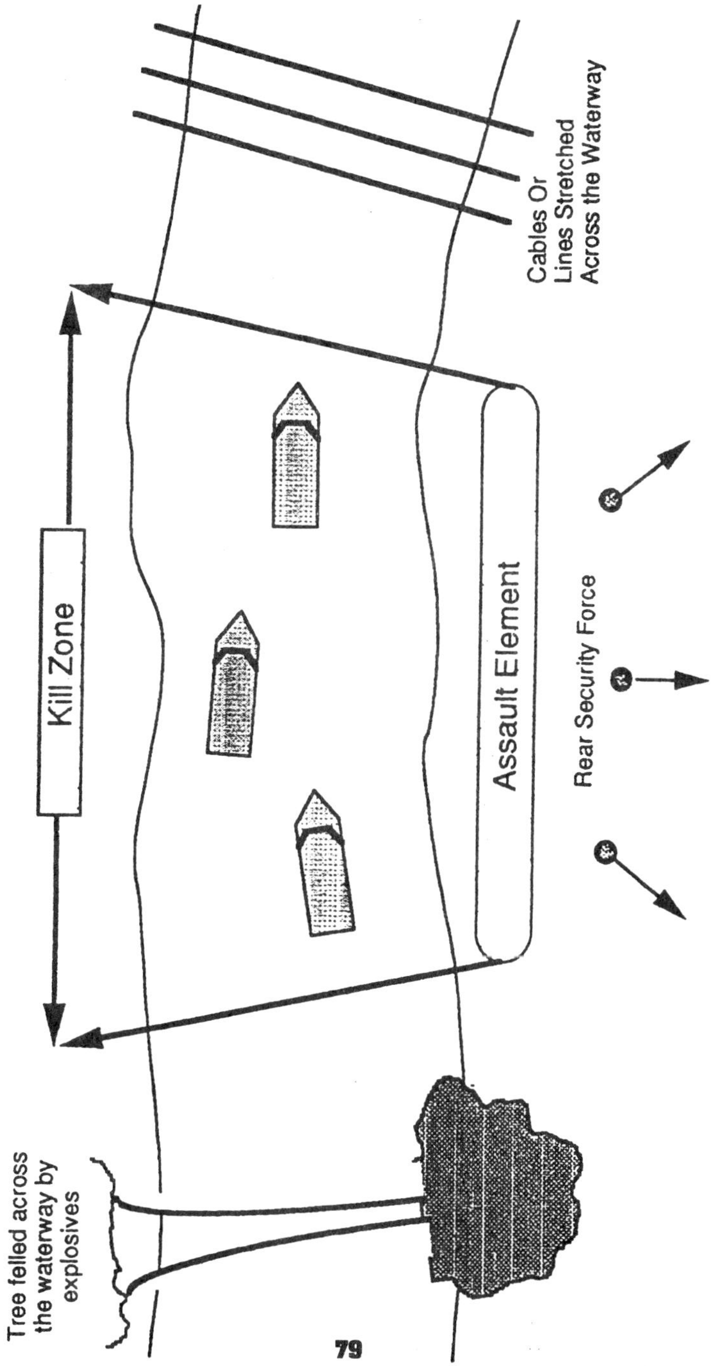

Kill Zone

Assault Element

Rear Security Force

Cables Or
Lines Stretched
Across the Waterway

Tree felled across
the waterway by
explosives

beads so that a boat or boat unit no sooner fights its way through one ambush and starts to tend to the casualties than it is hit a second, third, or fourth time.

During river operations, patrol craft operating with troops aboard are often split into two equal groups with an interval between groups of 300 to 700 yards. The actual interval will depend on the course of the river, terrain features, and whether it is a night or day mission (i.e., at night the interval may be reduced to 300 yards to ensure that groups can usefully support each other). When ambushed, the first patrol craft group will go through the enemy fire and beach the boats to land troops above the ambush site. The second group of boats will not run the ambush but will beach and disembark troops below the ambush position. Troops can then move, pincer fashion, on the enemy while patrol craft provide mortar support. Troops can be used to cut off the enemy's retreat by deploying them behind the ambush position.

In Vietnam, breaking contact with the enemy when being ambushed in patrol boats was a frequent experience for the U.S. Navy during their riverine operations. The tactic of laying down an immediate heavy base of fire from every weapon onboard and accelerating out of the kill zone was found to be most effective, except when using a heavily armored craft such as the Mighty Mo. In that case, it was often just as expedient to remain in the zone and fight it out with the heavy weapons on board. In smaller craft, it was found that employment of port and starboard 7.62mm mini-guns with very high rates of fire was effective in forcing the heads of the ambushers down, allowing the patrol craft to evade out of the kill zone.

Patrol craft on river operations without troops embarked can use the same split group tactic, with variations. The boat or group running the ambush can sail through and beach above the ambush position, but on the opposite bank. The second patrol craft group can do the same below the ambush. Both groups can then mortar the enemy position at will, without worrying about hitting their own boats. The mortar barrage can serve two useful purposes. It lets the ambusher know each time he hits the patrol craft he will be hit back—hard. It may also hold the attacker in the area until a helo or fixed-wing air strike can be called in to finish the job. It is important this action be taken after every ambush.

If the boats have seriously wounded personnel on board, these may be sent out of the river on one of the patrol craft while the others lay down mortar fire. If there are dead personnel on board, with no wounded, there is no reason to leave the river in haste. The mortar barrage should be conducted by all boats in that case.

Sometimes riverine ambushers set up preaimed firing trays in which a rocket launcher may be placed for firing at boats in the river. These trays are often used in conjunction with aiming stakes placed near the riverbanks. For this reason, an enemy's first salvo is often very accurate. However, once a heavy volume of fire from the patrol craft is begun, the accuracy of the enemy fire will be reduced considerably.

Mines are a constant threat in river or swamp areas. They may constitute a demolition ambush in their own right, or they may be used as part of a full-blown bushwhacking. Taking precautions against mines is central to counterambush techniques. Yet watching for telltale indications of mines in or near the water and searching the banks and areas near the water for indications of an ambush tend to keep a boat crew more than occupied. Often the crew forgets to look up. In areas of dense foliage, where there is growth that overhangs the waterway, claymore-type mines can be strung on trees, jutting out over the water. Three to five claymores, affixed to branches and pointing down to the water, can create a deadly explosive ambush all by themselves. These overhead attacks can be devastating because the limited defensive armor of most boats is designed to stop gunfire coming in horizontally. A series of claymore bursts overhead will mow down virtually all personnel on deck like a dose of grapeshot. When gunfire is added to the equation, casualties on the boat may be very high.

Boat commanders also need to be wary of baited ambushes, where a body, weapon, or small boat is left exposed on or near the bank. Sometimes this bait is held in place against a tree root as if it were wedged there by the current. When the patrolling craft noses up to investigate, the ambushers initiate the attack at close range. This tactic is especially useful in getting boats to pass beneath an overhanging tree or similar foliage where downward-directed claymore mines are emplaced. The tactic was used with deadly effect by the Vietcong.

Some rules of thumb proven out in Vietnam for avoiding ambushes of small patrol craft on rivers were the following:

- Use predawn cover of darkness for transit to the area of operations (AO).
- Avoid an established pattern of commencing operations at a particular time.
- Never sail into waters where the boat cannot be turned 180 degrees and would have to back down if caught in an ambush or stopped by a blockade of mines, tree trunks, etc., in the river.
- Avoid telltale use of aircraft directly overhead.
- Employ a feint toward the mouth of one river and follow up with attacks of likely ambush positions into which enemy ambush parties may have been drawn.
- Remove barricades and fish traps that restrict patrol craft passage or could harbor underwater mines.
- Employ coded checkpoints for reporting position of units; enforce strict communications security. Minimize voice traffic.
- Use at least two-thirds speed where possible—but not necessarily maximum speed at all times. By using

two-thirds speed most of the time, the boat can shift into higher speed immediately if waterway conditions allow.

Ambushes cannot always be avoided, but their effect can be neutralized by proper procedures. Reconnaissance by fire from a boat or patrol craft, using both automatic weapons and fragmentation projectiles, can pretrigger an ambush. Not only riverbanks, but trees and promontory landmarks also should be taken under fire to eliminate lookouts and snipers.

Weaponry on a small river craft is generally useful in most situations, but experience has shown that there are limitations to its use. Those limitations must be understood and planned for.

For instance, no matter what formation riverine patrol craft adopt as the boats sail along, nearly always some of the weapons will have their assigned fields of fire directed away from the target or masked by the patrol craft's structure. Other friendly patrol craft may also be between the craft and the ambushing force, masking return fire. In an engagement, the commander of the ambushed force has to maneuver quickly, keeping several ends in mind:

- The maneuver should bring maximum firepower against the ambushers.
- The maneuver should present the smallest possible target.
- The maneuver should prevent or minimize the chances of firing on, or being fired on by, friendly forces.

Riverine patrol craft, particularly small ones, are inherently unstable weapons platforms. Currents, obstacles in the waterway, the speed of the patrol craft, other craft, wave levels, etc., make stability of the craft, which is necessary for accurate firing of the weapons, questionable.

The weapons themselves may present problems. In many riverine ambush situations, the use of the .50-cal machine gun is severely restricted. Since the ammunition is not point-detonating, ricochets are a hazard to civilians and friendly troops in the area. In Vietnam, the dense civilian population of the Delta virtually eliminated the use of the .50 cal. for other than emergency situations. The 40mm cannon showed itself to be potent for both point and area fire using high explosive (HE) ammunition. Use of the 81mm mortar in both direct and indirect fire became common in Vietnam. Events proved that, when used against bunkers, a delay fuze should be used for maximum penetration. There was also no question that a mortar round that landed in a rice paddy or swamp was ineffective. The effectiveness of mortar fire against targets in swamps and paddies was increased when a time fuze was used for an 18-to-20 foot air burst. Recoilless rifles, bazookas, or flamethrowers were needed to respond to ambushes from riverside bunkers in order to achieve bunker neutralization. Using hull-mounted claymore mines as a counterambush weapon was sometimes useful.

Although most weapons have definite limitations and some are severely limited, mini-guns, with their high rate of fire, are almost guaranteed to break up any ambush.

It is also important to keep in mind that, at low tide, riverbanks occasionally are above the level of most guns installed on the riverine assault craft. Guns that normally would be used to respond to an assault may be useless at low tide or in areas where the banks are high. In that case, the attackers may be able to fire down into the boats, but the boats' weapons cannot be elevated enough to fire back.

Helicopter gunships can play an important role in preventing riverine ambushes. A helicopter fire team escort, one that is in direct communication with the commander of the boat unit, is useful in discovering and destroying ambushes. Artillery support of boats in transit can also be highly effective. In the event artillery support is available, the artillery forward observer should ride in one of the lead boats if there are several craft.

CAMOUFLAGE

In no other operation are camouflage and camouflage discipline more important to the attacker than before and during the ambush. New technology is changing this field faster than any book could keep up with. At present, for instance, thermal imaging and night sights/night vision goggles remove darkness as one of the major cloaks used by ambushers—and anti-ambush teams. The present off-the-shelf technology is a consideration even today. In the near future, new technologies will make camouflage even more of a problem and thus ever more important to the conduct of a successful ambush.

For the time being, however, most ambushers and counterambushers aren't likely to have the high-tech materials available. An ambush can still be set up successfully without the high-tech gimmicks. But it must be complete, all-around camouflage—360/180 is the shorthand way to say it. It must be adequate and cover 360 degrees in case the enemy approaches from an unexpected direction. Because of aircraft and intelligence-gathering drones, the camouflage must be effective from horizon to horizon and straight overhead—180 degrees.

Camouflage has significant importance in both military and civilian ambush attacks. Sticking to the basics will get an ambusher a long way, even today. And the most basic lesson is that the best camouflage is, where possible, to have weapons fire through screens of live, undisturbed vegetation. The idea of camouflage is to mislead potential enemies regarding position, presence, or identity. A majority of all targets on the ambush field are seen because of either the lack of or improper use of camouflage. Camouflage is not only essential in setting an ambush, it can be a key factor in efforts to foil an ambush as well, particularly once the target is able to maneuver.

Camouflage has been less a necessity for the ambushed party than the ambushers in most cases, and that is particularly true in ambushes where the target is a civilian or protected person, as in an executive ambush.

Camouflage really means to fit in with surroundings. For instance, a potential civilian terrorist target should have a vehicle painted in the most popular but conservative color used locally. A Mercedes painted in urban military camouflage markings, with chrome removed to reduce shine, would actually stand out on the street; so would a cherry red BMW. In most communities a silver or black car would blend in, making it more difficult for an attacker to be certain whether he, she, or they had the right target. Camouflage in the urban terrorism environment is more involved with things like avoiding distinctive markings, clothing patterns, or vanity license plates on vehicles than it is with greasepaint and multicolored cloth.

In military-style ambushes, the three key visual camouflage factors are: *shine, regularity of outline*, and *contrast with background.*

- *Shine.* Any object that reflects light should be camouflaged. Virtually all metal objects, not to mention many plastics, produce some sort of shine at certain angles.
- *Regularity of outline.* Humans, vehicles, weapons, and helmets are easily recognized. For that reason, camouflage is necessary to break up the shape of these objects.
- *Contrast with background.* When choosing a position for concealment, try to select a background that visually "absorbs" the person and equipment. That may seem self-evident, but there is also an important foreground factor to consider. A parapet of exposed new earth thrown up around a machine gun position will tip off an ambush target as quickly as waving a weapon and whistling.

Consideration must be given to many factors, not all of them apparent, when dealing with contrast. Take the use of colored parachute cloth, for instance. In daylight it is an excellent camouflage material. At night the cloth shows a distinctive signature when the area is viewed through infrared goggles. The cloth, rather than hiding the position, can actually draw attention to it. Of course, if the potential target doesn't have infrared night vision goggles, the use of parachute cloth doesn't make any difference. However, if the target does possess such equipment, a thoughtful commander has to weigh the chances of being seen *because* of the fabric's signature against the utility of the cloth for concealment during the day and at night when high-tech equipment is not being used.

The subject of camouflage is a complete study in itself, and a short section such as this cannot adequately cover the field. However there are some key things to remember. Badly cut lanes of fire—ones that display an eye-visible regularity, for instance—are nothing so much as a big finger pointing straight at the ambusher. So is natural camouflage material, such as branches, where the leaves have begun to wilt. Often the differing color and texture of the wilting vegetation can be seen from dozens of yards away.

Noise and light discipline are another form of camouflage. Without noise and light discipline, the most elaborate ambush may founder. The glow of a cigarette, for instance, is visible for hundreds of yards. On missions—and this applies to both ambushers and potential ambush targets—smoking is dangerous to your health.

In all too many cases, camouflage is designed to be effective at eye level only. But in any potential ambush situation, it is important to consider that the enemy may be able to get a different angle on things. Observation from nearby hills, a tall structure, aircraft, or even treetop posts can negate the most effective eye-level camouflage if precautions haven't been taken against overhead observation. The best bushwhackers know this.

There are many readily available camouflage materials. Stick-tube camouflage paint is effective in covering exposed parts of the body—face, neck, and hands. When applying stick-tube paint, two people should team together to make certain the coverage is complete and effective. A signal mirror will help, but it cannot match the results obtainable when two people help one another. Using two persons, for instance, can be much more effective on the back of the neck and other hard-to-see places.

The color of the camouflage paint will depend on the terrain and environment in which the operation will be carried out. The forest green, olive drab, and black used in jungle camouflage would be worse than useless in Arctic environments or desert conditions.

When camouflage paint is unavailable, field expedients come into play. The most common field expedients are *mud, cloth,* and *foliage.*

• *Mud.* It should *never* be used on the skin. The high bacterial content makes it a health hazard, and it runs off when perspiration or precipitation hits it. Mud is best used to camouflage shiny objects, including belt buckles.

• *Cloth.* It is often used to break up the line of weapons and people. In attaching a garnish of cloth to the operating uniform, the pieces of cloth should be sewn loosely so that they overlap and create an irregular pattern of line, texture, and color. Cloth can be colored with mud, charcoal, burned cork, or coffee grounds. The detectable odors of fuels, oils, and greases make them unsuitable for coloring cloth in most cases.

• *Foliage.* Natural foliage is usually preferred over artificial camouflage materials but often is difficult to affix to the gear or body. Rubber bands, including makeshift ones cut from discarded inner tubes, are often effective in securing foliage. Remember to replace natural foliage when it starts to wilt. If using foliage as camouflage while moving up to an ambush position, be aware of changes in the foliage pattern and adjust accordingly. Moving "pine trees" into an oak grove attracts the eye, for instance.

Modern technology has changed the meaning of

camouflage in some cases. Night vision goggles and infrared sensing devices have, for years, canceled darkness from military equations where they were used. The even newer thermal imaging sensors—which use the heat of the human body to create a picture—promise to make much more drastic changes in the way nighttime ambushes are planned and carried out. High-tech materials that do everything from dispersing laser sight spots to eliminating the "return" on battlefield radar signals are new camouflage techniques. But at present and in the foreseeable future, this kind of technology will have an effect only on ambushes involving Special Operations forces and others with access to such closely guarded material.

Most ambushers or ambush targets don't have to take such high-tech camouflage/countercamouflage into account as yet, however. The list of key considerations that should be taken into account when doing camouflage at an ambush site is still pretty short.

- It is never wise to provide the ambushed party with target indicators—spoiled, browned, or wilted foliage; cover or concealment that is not native to the area.
- While preparing a camouflaged position, some members of the unit must act as lookouts while the others prepare the position.
- Where possible, after the primary ambush positions are built and camouflaged, alternate positions should be created—ones that are accessible by a covered route.
- After a prepared position is created, it is important to inspect it from various angles to make certain that it does, in fact, meet the requirement of concealment. Shine, regularity of outline, contrast with background, and noise and light problems should have been fixed.

A final note on camouflage, which is essentially concealment. Never mistake it for cover! Camouflage is in no way protection from hostile fire. Camouflage only makes it difficult for the other party—whether that is the ambushed or the ambusher—to know where to fire.

CANVAS COVERS

Canvas covers on military trucks should generally be removed in potential ambush situations. While they give concealment to the troops inside, they make observation by the troops impossible and hinder their reaction. Troops in a covered vehicle end up taking rounds without being able to see where the fire is coming from. Because of that they lack the ability to return the high volume of fire needed to break up an ambush.

CAR BOMBS

Some car bomb attacks have characteristics of a raid rather than an ambush, but nearly all have at least some elements of a bushwhacking.

There are two types of car bomb ambushes. Although they may be employed against military-related targets, most are generally used against executives or protected people of some type, people traveling alone in a single vehicle or a very small convoy.

The Ulster Car Bomb is a terrorist tactic in which an explosive charge is planted on or underneath the vehicle. Sometimes it is detonated by a triggering action of someone in the car—anything from opening the door to get in to turning on the ignition, shifting gears, or driving a set number of miles. At other times, this type of car bomb is detonated by remote control. The Ulster Car Bomb attack has been very popular with the Irish Republican Army (IRA) for eliminating individuals. The IRA has even adapted it for use against people starting boats and used it to kill the British Commandos' original commander—Lord Mountbatten. (The IRA also uses another car bomb attack that is more in line with a raid, leaving the vehicle at a fixed target. The two should not be confused.)

The Lebanese Car Bomb can also be a true ambush. A car, truck, cycle, or van packed with explosives is left at some location. When a car or convoy carrying the ambush target rolls past, the explosives are set off by remote control. Other variations of this tactic involve timed, rather than remote-controlled, ignition to create random terror. The ambush "targets" are whoever happens to be in the location at the time.

Yet another variation uses a suicide bomber. The suicide bomber variation has been used with good effect by Shiite Muslim groups against Israeli military convoys; in its "raid" form it was used with deadly effect against U.S. and French targets in Lebanon. The Liberation Tigers of Tamil Eelam also have a history of using the suicide car bomb/cycle bomb with deadly effect.

CLANDESTINE INSERTION OF AMBUSHERS

The clandestine insertion of ambush forces into an area can be accomplished by moving them into the area as part of a regular military patrol. The ambush elements should be dispersed throughout the patrol formation with their radio antennas detached. Upon completion of the ambush, another patrol can be used to pick up the original ambush force and drop off another, if desired. In this way the patrol size is kept constant, making it difficult for the enemy to notice that an element has been dropped off.

CLEARING VEGETATION

In guerrilla-infested areas, vegetation may be completely cleared from roadsides to ranges traditionally used by the attackers—in the case of the VC, that was about 90 yards. Manpower requirements may limit the clearing of vegetation to particularly dangerous stretches of roadway.

COLUMNS PROTECTED BY ARMOR

Attacks against columns protected by armored vehicles depend upon the type and location of armored vehicles in the column and the weapons in the hands of the ambush forces. From an ambusher's position, where possible, armored vehicles are destroyed or disabled by fire of antitank weapons, land mines, or Molotov cocktails, or by tossing grenades into open hatches. An effort should be made to immobilize armored vehicles at a point where they are a) unable to give protection to the rest of the convoy and b) block the route of other supporting vehicles.

COMFORTABLE WAITING

Comfort, or rather discomfort, is a major factor in preparing a successful ambush. For that reason all members of the ambush team should answer the call of nature before getting into position. They should also drink plenty of fluid before doing so; it is important to keep fluid levels up in the body, and by drinking fluids before going into position it is less likely the ambushers' positions will be given away by someone making metallic, sloshing, or gurgling sounds later. It is often a good idea for members of the ambush team to wear more or heavier clothing than they would on a march. The clothing should cover as much skin as possible as a protection against insect bites.

Since movement is reduced to virtually nothing at an ambush site, the warmth will be tolerable, if not welcome. When in position, all movements should be rationed. But every quarter or half an hour, at least one part of the body should be moved to stave off numbness and cramping. Before making even the most minor movement it is important to look to see that the part that is going to be moved will not dislodge anything, make noise, or visually alert an approaching enemy. Then, any movement should be slow and deliberate.

CONTROL

In conducting an ambush, control involves a number of key factors. Control factors have to be carefully built into the ambush plan. Proper control means:

A. There will be early warning of the approach of the target.
B. Fire will be withheld until the target is inside the killing zone.
C. Appropriate steps will be taken if the ambush is detected prematurely.
D. Withdrawal to an easily recognized rally point will be done on time and efficiently.

CONVOY SECURITY DETACHMENTS

On roads through hostile areas or in areas where ambush of even civilians is a possibility, lone vehicles and civilian convoys not capable of providing their own security are grouped and escorted through dangerous sections by armed security detachments. All traffic through danger areas is controlled by traffic-control stations.

Convoy security detachments are specially organized and trained to protect convoys from ambushers. They have adequate fighting power to counter any likely attack. A detachment may be organized into two or more parts: a holding or defending element and an attacking element. The size and composition of the force will vary with the geography, the capabilities of the likely ambushers, and the size and composition of the convoy. An armored infantry company is well-suited for this work.

A typical security detachment might be organized in the following way:

1. The headquarters detachment provides the staff, communication, and medical facilities.
2. The armored element provides increased firepower and shock action.
3. The infantry detachment may be organized into a holding element and an attacking element.
4. The combat engineers supplement the holding element and are used to make minor bridge and roadbed repairs. They may also be used to detect and remove mines.

Before entering the danger area, the convoy command responsibility is clearly fixed. The commander is briefed with the latest information about the area to be traversed. He draws up his plans and issues orders that include data on the convoy formation, intervals between echelons and vehicles, speed of travel, and detailed reaction plans if an ambush is encountered. All elements are briefed to act initially according to prearranged plans, as there will seldom be time for a warning order to be issued on the road.

The canvas covers on trucks are removed and tailgates are left down. When practicable, personnel are placed in vehicles so that they can detruck rapidly. Arms and weapons are readied for immediate action and senior personnel in each vehicle are made responsible for seeing that all passengers are on alert when passing through danger areas.

The formation of a security detachment and its integration into a convoy may, and should, be varied. Guerrillas may be expected to observe convoy habits and will prepare their ambushes to cope with expected formations.

The holding element of the security detachment is distributed to provide close-in defense throughout the convoy.

Armor generally leads the convoy as a defense against mines. When armor is not available, a heavy vehicle with sandbags placed on the floor beneath personnel should lead the convoy.

Any prisoners being transported may also be placed in the leading vehicle, though in wartime, strict adherence to Geneva Convention rules regarding the use of hostages and the requirement to protect prisoners from harm must be observed.

Any remaining armor is distributed in depth throughout the column to strengthen the defense of the formation and to provide supporting fire for the attacking elements. Armor's radio net also provides a ready means of communication throughout the convoy.

Sandbags should also be placed along the sides of troop-carrying trucks as a defense against small-arms fire; that can be backed by sheet iron. Essentially, this makes the truck a moving foxhole.

Convoys may be escorted by reconnaissance aircraft, either fixed- or rotary-wing. Strike aircraft should be on call, and where possible, artillery should be available.

Where ambush is likely, the speed should be kept slow—about 10 to 15 MPH. When passing through areas where ambush is most likely, such as areas overgrown with heavy underbrush, reconnaissance by fire may be used to keep the heads of potential attackers down. Firing is allowed, however, only on the orders of the convoy commander.

Advance guards on a convoy are generally ineffective against ambushes since the attackers will usually allow the advance guard to pass the site of the main ambush, then block the road and deal with the main body and advance guard separately. However an advance guard can be useful if it is thought of as a quick-reaction reinforcement team that can attack the flanks or rear of the ambushers.

At the first indication that an ambush has been set, vehicles stop. The drivers take care to remain in the tracks of the vehicle ahead of them. No effort should be made to get clear of the road by driving to its side or onto the shoulder. It is best to always assume those are mined.

Personnel, other than drivers and assistant drivers, detruck as rapidly as possible without waiting for the vehicles to come to a complete stop. The drivers turn off ignitions, brake their vehicles to a halt, set hand brakes, and leave the vehicle in gear before disembarking the truck. Assistant drivers are alert to help if the driver becomes a casualty.

Upon disembarkation, personnel take cover and open vigorous fire on suspected targets. Tanks open fire as they maneuver to the most favorable ground in the immediate vicinity. If the convoy is in radio contact with other friendly forces, the ambush should be reported immediately.

The security detachment commander, after quickly surveying the situation, issues orders to the commander of the attacking element to begin one of the prearranged attacks, preferably an envelopment. The fire of the holding force is coordinated with attacking elements by prearranged communication.

After driving off the ambushers, security details are posted to cover the reorganization of the convoy.

Ambushers captured in the action are interrogated about the rendezvous point where they were to reassemble. At the earliest opportunity, the convoy commander reports by radio to a road control station, giving a brief account of the engagement and such information as may have been secured from captured ambushers.

Where practicable after an ambush, patrols are sent to apprehend, interrogate, and take any appropriate action against civilians living near or along the routes of approach to the ambush position. This can seldom be done by the convoy security detachment without unduly delaying the arrival of the convoy at its destination and subjecting it to further ambushes. This action, therefore, is normally assigned to other security troops stationed nearby.

In convoys where there is no formal security detachment but troops are present, the preferred scenario is a little different. Under these circumstances, part of the troops are placed well forward in the convoy, and a strong detachment is placed in a vehicle that follows the main body by about three minutes. Radio contact is established between the two groups if possible. Fairly fast speed is maintained when road conditions permit, especially through defiles. Advance troops on foot reconnoiter sharply curving roads, steep grades, or other areas where fast speed is impossible.

At the first indication of ambush while the convoy is in motion, leading vehicles—if the road appears clear—increase speed to the maximum consistent with safety. They attempt to smash through the ambush area. Drivers or assistant drivers of vehicles disabled by ambushers' fire or mines try to get their vehicles to the sides or off the road so vehicles in the rear can continue to get through.

Troops from any vehicles stopped in the ambush area dismount and return fire. Troops from vehicles breaking through the ambush area dismount and attack back against the flank of the ambush position. The rear guard of the convoy, upon learning that the main body has been ambushed, dismounts and attacks forward against the flank of the ambush position. Attacking groups must take care that they do not attack each other!

If the ambushers allow the main convoy to pass through and then ambush the rear guard, troops from the main body return and relieve the rear guard by an attack against the flank of the ambush.

DAY-NIGHT AMBUSHES

One tactic that came from Vietnam was the day-night double-punch ambush. Sometimes after a night ambush had been sprung, but often because the VC had good intelligence and knew an ambush had been set for them and avoided contact, the ambushing party would leave. The VC would sometimes search the area where the Americans had been in hopes of scavenging material that might have been left behind. Some cagey U.S. commanders would leave a daylight ambush in a position that had been occupied during the night, waiting for the searchers to show up.

DEADLY SINS

Some very common tactical errors will get troops decimated in an ambush. They include:

1. *Troops not alert and not carrying weapons ready to fight.* Signs of this are soldiers strolling along, unaware of anything taking place around them; weapons being carried on the shoulder or slung across the chest; setting sights at zero clicks elevation; and sometimes even suspending items from either or both ends of the weapon. To solve the problem, soldiers must be fully briefed on the operation and latest intelligence. They must be assigned a sector of observation and fire, and frequent checks should be made to ensure they are alert. Weapons must be carried at the ready, with sights properly set on their battlefield zero. Rations and other items should be placed in, or tied to, the combat pack. Inspections must be held to ensure that only essential items are carried.

2. *Troops not maintaining proper intervals.* This causes numerous problems. When soldiers move along in bunches, three to six feet apart, most cannot fire their weapons without hitting their own troops. A mine or grenade detonation causes multiple casualties. It is impossible to identify a platoon or squad within a company, as intervals are not maintained between subelements. Solving the problems means that during daylight, individual soldiers must maintain a proper interval—usually about 30 feet. They must have room to observe

their assigned sector and be able to fire their weapons without endangering friendly troops. Between squads, 70 to 100 feet should be maintained; at least 175 feet is needed between platoons. During periods of reduced visibility and darkness, the interval between individuals and units may be reduced.

3. *All-around security not established on the move.* When a company or battalion moves on a road without a security element to the front, rear, and flanks (as is frequently done), no warning of any ambushers is possible. Therefore, when contact is made with the enemy, the main body is immediately brought under fire and usually pinned down. To prevent this, companies and battalions must have elements posted to the front and rear and at each flank. Security elements must check every area within small-arms range of the main body that could conceal an enemy force. When an ambush is observed, the main body must be warned by radio, hand and arm signals, or warning shots. Security elements must immediately place fire on the enemy, while elements of the main body maneuver to assault.

DEMOLITION AMBUSHES

Demolition ambushes may be either deliberate or opportunistic. They can be classified as point or area ambushes, as well. Many feel there is a fine line between a demolition ambush and some other form of ambush where explosives are used heavily. In the purest sense, however, a demolition ambush is one in which there is no "assault" element other than those triggering the explosives. In cases where there is an assault element, demolition personnel are usually considered to be a part of the support element.

Demolition ambushes are particularly useful when the objective is the destruction of one part of a convoy or column or a single vehicle is being targeted.

The following are special considerations in planning for a demolition ambush:

1. When selecting terrain, choose an area along a path, trail, or road that is bordered by woods, brush, swamp, cuts, or water. Place the ambush on a hill or curve if possible. When negotiating hills or curves, moving elements are slowed down, making them more vulnerable to fire of all kinds and reducing the problems associated with timing the triggering of explosives.
2. Acquiring necessary information about the enemy prior to selecting the site is important. The essential elements of information (EEIs) in this case are time, terrain, and movement of target.
3. Problems and issues involved in the construction of mines, fragmentation charges, and demolitions (main and branch lines).
4. Problems and issues associated with the placement of mines and/or charges. (The number of mines to employ is dependent upon size of target.)

Demolition ambush, as used here, is the formal military name. When used by terrorists and irregulars, this tactic goes by several other names (see also Roadside Bomb Ambush, Lebanese Car Bomb Ambush, Executive Ambush, and Armored Vehicles).

DESERT AMBUSH SPECIAL CONSIDERATIONS

In a desert environment, the likelihood of encountering vehicles increases. Because of the unique problems associated with stopping vehicles and attacking troops who are using vehicles as cover, extra rounds should be carried by all members of the ambush team. Rockets and antivehicle weapons are also needed. It is particularly important, in the desert, to enter positions through the rear of the ambush site. In desert terrain, evidence of the ambush team's movement cannot be easily removed. The sand/dirt surface is easily marked, and erasing the signs is nearly impossible. A lack of precipitation or foliage that would otherwise remove or cover some of the signs of the ambushers' presence makes it critical not to enter the killing zone at all.

DIESEL-POWERED VEHICLES

Diesel-engined vehicles, whether military trucks or civilian cars, are preferred when dealing with potential ambush situations. Diesel fuel is significantly less volatile, and a puncture of the fuel tank generally won't produce a fire. Gasoline, being a low-grade explosive, poses a significant risk—particularly when mines, roadside bombs, or rockets are being used against the vehicle. The thermal effects of a gasoline explosion may cause a fatal fire in a situation that would otherwise be survivable.

DISTANCES

There is no firm rule regarding the optimum distance between the ambushers and the target. Everything depends on cover and concealment available, and the marksmanship and accuracy of the ambusher. The VC generally opened fire on vehicle columns from a distance of 9 to 60 yards, although in rare instances the VC took advantages of available cover and opened fire within distances as short as one or two yards of the roadway. In roadside bomb and explosive ambushes the term "distance" has no meaning, as these are generally stand-off attacks using remotely detonated explosives. Crude bombs made with homemade explosives may need to be either close to the target or incredibly large—so large it takes a truck to haul the stuff. A crude bomb designed to set off gas canisters may be of moderate size. On the other hand, an expertly made bomb using military explosives may be quite small and yet be able to do damage or kill from long distances.

EXECUTIVE AMBUSHES

Business leaders, political officials, and community figures often become the targets of terrorists or other irregulars who have a "fist in the mouth" attitude toward politics and economics. These attacks are usually designed as either assassination ambushes or prisoner ambushes. Whoever the actual target—government official, business leader, community figure, or just the chance passerby whom terrorists decide to kill in order to spread fear throughout the community—the type of ambush is referred to as an executive ambush.

Ambushes directed against executives and protected personnel are usually somewhat different than ambushes involving military personnel. Civilians generally can't plan for, or respond to, an ambush in the same way that a military force does.

The military leader has a great deal of flexibility in many cases as to when a convoy will start moving. Time schedules can be devised—and timing is a factor that should always be considered—to deter ambushes. Secrecy can be enforced, when necessary, by isolating the military troops before they jump off.

With executives, the situation is entirely different. They are a single target. Often they have no protection except their wits, knowledge, and instinct. They run on a schedule. Unpredictability, a key to anti-ambush planning, is usually not an option for executives. Executives have appointments. They meet clients, co-workers, and other executives. They show up places on time. They have office hours. There is seldom much day-to-day leeway on when the executive will leave for the office. It may be 9 A.M. one day, 8 the next, and 8:30 the following day—but even that broad a span would be unusual. In most cases, executives, who are generally more focused on running their business or operation than they are on staying alive, will establish a daily movement pattern that has no more than five minutes of difference over an entire work week.

Since meetings and appointments are scheduled in advance, anyone who has access to that information, or can get access to it, can determine almost down to a few seconds where a potential target will be. A cleaning person in the executive's office, with access to a daily calendar, can pinpoint the whereabouts of an executive for terrorists! Varying the time significantly is seldom possible, and the advice is too often based on naivete.

The "vary your route" advice that most security management people offer is also good, but equally naive. In all too many cases, the route an executive can use is a heads-tails choice. He can go right or he can go left out of his driveway. It is extremely difficult to change routes near either the home or office. As a result, the further from home an executive gets, the less likelihood there is that an ambush will take place—until the executive nears the office. Then the chances increase again because, in the final analysis, he ends up with the same two choices—left or right. And sometimes an executive doesn't even have that much choice. One-way streets can further restrict the possible directions.

Terrorists have learned that the best time to stage an ambush against an executive is when the target is on the way to work—and in most cases that is in the morning. In fact, few executive ambushes take place at any other time! Terrorists have also learned that the best place to stage the ambush is at or near the home or in some cases the office. (See also Leaving the House.)

Usually, however, the actual physical attacks are preceded by days or weeks of surveillance. Spotting the surveillance will probably lead to breaking up the attack. That's why, while driving, it is important to check often to see if there is a tail. This is the first indication that an attack is planned. When there is any suspicion that a tail is in place, it's important to get as complete and accurate a description of the vehicle and its occupants as possible. The information should be given to security officials.

In most cases, terrorists attempt to stop the car at or before the moment of attack. The best, and often only, defense is to keep driving. Moving targets are harder to hit accurately, and it's impossible to kidnap a person whom you can't physically get to. Since executive ambushes usually involve either assassinations—in which the standard pattern is to use automatic weapons to rake the car and any occupants—or kidnappings, the best defense is to keep moving at all costs.

Terrorists know that, so they'll use roadblocks—anything from semitrailer trucks parked across a street to baby carriages. In most cases the vehicle will be blocked from the front and, in many cases, the rear. It's at this moment that the driver needs to employ evasive driving techniques—techniques learned long before and practiced often.

In an ambush, passengers should get down and take cover. Passengers can perform a service by calling for help over the car's cellular phone or radio. The driver, using the rear-view mirror, has to make a quick determination whether to go forward or back. The rule of thumb is to go forward if at all possible since the vehicle will generally be in the killing zone for a shorter period of time.

Evasive driving, a subject so complex that textbooks are written on it, can be used to turn the vehicle or ram through the roadblock. (See also Evasive Maneuvers.)

Lebanese Car Bomb ambushes may also be used to kill the occupants of a vehicle. Frankly, there is precious little defense against this or the roadside bomb. The car and driver will either survive or not.

The chances of survival in a car bomb attack can be improved by speeding up or slowing down at spots where a car bomb or roadside bomb attack is most likely. A sudden change in speed, in many cases, will throw off the timing of someone operating remotely detonated explosives, resulting in the target vehicle being either ahead of or behind the killing zone. It's important to note that a car standing at a stop sign or a red light is absolutely stationary. In that case, the car's speed and position are canceled from the equation; the vehicle and driver are at the mercy of the ambusher and luck—good or bad.

The driver who survives a car bomb ambush can only keep driving and call for assistance. Never stop to see what happened or who needs help. That only gives the ambushers a second shot—and while they might not have a second arrow in their quiver, there is no sense in finding out for certain.

EVASIVE MANEUVERS

In an executive ambush situation, the key to survival is to keep moving and take evasive, even aggressive steps. The subject of evasive maneuvers is a specialized field and cannot be developed at length here. In short, evasive maneuvers work to break up an ambush because the target's vehicle is a poorer target, it can be used as a weapon, and the ambushers aren't properly prepared to deal with a skilled driver.

Ramming, turning, and *positioning* are three basic classes of maneuvers. All of them are highly dangerous and should be attempted only after training with professional instructors.

Ramming

Terrorists will often put up a roadblock in front of a target vehicle to stop it in the killing zone. Usually that roadblock is some other vehicle. Drivers, who are taught from the day they first pick up a car key to avoid crashes, instinctively shy away from contact with another vehicle. They stop dead rather than ram. But often the best possible move in an ambush is to use the car's weight, speed, and power to force a way through the killing zone. Doing so generally thwarts the ambushers; very few have contingencies for continuing the attack on a vehicle that escapes the killing zone.

Modern vehicles, particularly cars and trucks, will take a great deal of punishment. They can ram their way through most roadblocks. Obviously, when a car goes up against a semi-trailer truck, a heavy-duty dump truck, or a loaded cement mixer, it's pretty unequal combat. But a good evasive driver can usually punch his way through a one-car or two-car roadblock. The key to doing so is to force the blocking vehicles out of the path, while doing as little damage as possible to the ramming vehicle.

The best way is to strike the car being used as a roadblock squarely on a wheel with a fender, then power through in low gear. The idea is to bulldoze a way through, not smash through as movie stuntmen do. As a rule, aim for the end of the blocking vehicle that is lighter—the end without the engine. But it is always best to ram whatever end of the blocking car will give the most room for escape.

Keep accelerating after the impact. Sometimes the roadblocked car will get hung up on some bent chrome and be dragged a short distance, but if the driver of the targeted car keeps a foot on the accelerator, it won't be tailing along for long. It's important to drive and keep driving. As long as the car will run, it should be driven—flat tire or not.

In some cases, where it is clearly impossible to evade a roadblock or ram through it, a driver may opt to ram any relatively frail-looking building on either side of the street. If the car can be driven completely inside, the driver and passengers may have a chance to escape through the rear of the building on foot if not seriously injured. In any event, the crash will create confusion and much noise. In many cases, the ambushers will retreat because they haven't planned for such a scenario.

Turning

Turning during an attack allows the car to keep moving for longer distances and makes it more difficult for ambushers to draw a good bead on the ambushed car. Turning maneuvers allow a vehicle to use some direction other than straight ahead to escape. Classic turning maneuvers include the Bootlegger's Turn, a maneuver that will slew the car around 180 degrees in its own length, and the J-turn.

Not all vehicle ambushes involve barricades. In some cases, gunmen in another car may fire at the protected person's vehicle. Again, classic turning maneuvers are helpful to the driver who knows how to execute them.

But in a pinch, a simple hard turn—nothing fancy—may well work. Gunmen firing at a car will be leading it, expecting the driver to steer a straight course. Their own driver is expected to steer straight, allowing the shooters a steady aim. By moving left or right, or making a sudden turn, a targeted driver can deflect the terrorist's aim.

The vehicle in which the ambushed driver is riding may be the only weapon he or she has to fight back with. It should be thought of as a 4,000-pound projectile. But in fighting back a driver has to give up the ingrained tendency to avoid contact with another vehicle at all costs. That is why, in many cases, the smartest move a targeted driver can make is to attack the ambusher's vehicle with his own.

Motorcycle-mounted terrorists are particularly vulnerable to turns in front of them. A turn in front of an overtaking and passing motorcyclist leaves the riders the option of doing unequal combat with the car or breaking off their attack. Even a turn into the general path of the cyclist will send the cycle driver skittering away; the gunman, who is usually mounted pillion, has limited mobility under the best of circumstances. When forced to aim from a cycle that is veering and leaning at new and sometimes crazy angles, the gunman is likely to spray the surrounding countryside rather than the target vehicle with lead. And that is what breaking up an ambush is all about.

Hitting a pursuing or passing car with a bumper in the wheel-well area will often throw that vehicle out of control. The standard practice, when a gun attack is being mounted from passing cars or trucks, is to pump the brakes as the pursuing car comes alongside. That maximizes the difference in relative speed of the vehicles as they pass and plays havoc with the aim of the gunman. As the ambusher's car passes, a sharp turn into the other vehicle—aiming the car at the rear wheel well and axle—will cause the other vehicle to spin out of control.

Positioning

Positioning can be used to thwart certain types of ambushes, particularly those mounted from another car or a motorcycle. In the United States, forcing an ambush car to pass on the right rather than the left, for instance, means that gunfire from one of the attackers—one in the front—will be masked by the driver of the ambushing vehicle. Positioning the vehicle against a cycle-mounted gunman is perhaps the best way to break up the attack—and perhaps to break up the gunmen if you get really lucky.

FIRE AND MOVEMENT

This is a basic military method of assaulting or breaking contact. The teams leapfrog each other as they advance toward an objective, either directly or on a flank. They also use a leapfrog action in what is formally termed a "retrograde operation," something colloquially known as "getting the hell out of there."

Fire and movement sounds simple; it is infinitely more difficult to do correctly and safely. The technique requires effective fire and skillful movement to be successful. Because fire and movement is not instinctive, the technique must be continually rehearsed until the procedures become second nature.

In ambushes, fire and movement will generally be used by the ambushed group when assaulting the ambush team and attempting to disrupt the attack. Ambushers will often use it to break contact. Seldom, except in prisoner ambushes or resupply ambushes, will fire and movement be used offensively by the ambushers. Ambushers generally rely on surprise, defensive positions, and high volumes of effective fire in carrying out their attack, not fire and movement.

In practice the unit is broken into two groups. One group is designated to provide the base of fire while the movement team moves forward, backward, or to the flanks—depending on the orders from the unit commander. The two groups immediately establish a "dead space" between them, one that the fire team is not shooting into and one that the movement team is operating in.

Before making any move from one location to another, the members of the movement team should identify the next location for cover. It is important not to get caught in the open. Generally, the movement is on the run, but not always. Crawling may be the most effective means of movement in some cases.

The first team should move about half-way to the target—but the critical issue here is exposure time, not physical distance. It takes about 3 seconds for even a good marksman to acquire a target and fire. If the exposure time is limited to 2.5 seconds, the enemy will be left only with the option of firing blindly into a zone with automatic weapons or using area weapons such as flamethrowers and grenade launchers.

When the movement team reaches its position, or line, the members start to lay down their own curtain of fire. When the team that has been providing the base of fire hears the fire pick up from the movement element, the members know it is time for them to move, preferably about 3 to 5 yards past the team that moved previously.

The process is repeated until the enemy breaks contact or all members of the unit are out of the killing zone.

FLEXIBILITY OF TIME SCHEDULES

In military counterambush operations, assigned time schedules may become so inflexible they interfere with the mission. Ground tactical plans must be flexible and permit deviation from assigned schedules when the element of time conflicts with the mission. When possible, schedules should be determined by the senior commander on the ground.

Time schedules were a constant problem in Vietnam, and a government study noted that in one case during "a helicopter operation conducted by a Ranger Battalion of I Corps, one element was given the mission of searching and securing three objective areas within a specific period of time. During the operation it was found that sufficient time was not available to complete the mission so the commander decided to search only two of the areas. In another operation, a battalion of the 1st Division located VC trails but followed them only part way because the time allotted was not sufficient to adequately search the trails."

FREEDOM OF MOVEMENT

In a potential ambush situation, anyone who could be a target or who is protecting a target needs freedom of movement. Cars and vehicles should not be overloaded with personnel or equipment—a rule that applies equally to military convoys and the cars of protected persons. When vehicles are loaded to the maximum, people will be unable to use their weapons effectively. Protected persons will not have room to duck out of the line of fire. In a civilian security follow car, for instance, a fifth man in the car usually will just get in the way. In a 2 1/2 ton (6X) military truck, 16 or 18 men is probably a maximum number for effective response in an ambush situation.

HAMLET DEFENSE AMBUSHES

In Vietnam, small ambushes were established outside a hamlet to warn of the direction of an attack. These ambushes were located 1,500 to 3,000 feet from each corner or side of the hamlet and on likely avenues of enemy approach. These positions were varied and moved one or more times to preclude setting a pattern. In addition to these ambush positions, ambush defenses were planned inside the hamlet with a rehearsed course of action for each possible situation. Hamlet defense ambushes are a useful technique for local self-defense forces.

HARASSING AMBUSHES

If the purpose of the military ambush is to harass and demoralize the enemy, the ambush may not be carried out against the main body of a column, as is the more usual case. In this case the advance guard is targeted and fire from the assault element is directed at them. That forces the enemy to use disproportionately strong forces for advance guard duties. That may leave other portions of the column vulnerable or else force the enemy to divert additional troops to convoy duty. Such attacks have a decidedly negative psychological impact on enemy troops. The continued casualties among the advance guard make that duty unpopular for all but the suicide-prone.

HELICOPTERS IN THE COUNTERAMBUSH

In Vietnam, some counterinsurgency and counterambush operations went by the name "Eagle Flight

Operations." The fact that they failed to stem the tide and end VC and NVA ambushes is sometimes used to discredit this tactic. But those who take that view have an incredibly narrow picture of that particular war. Vietnam was a tangled web of political, social, economic, and military threads, a skein of tangled yarns. No single tactic, no individual stratagem could be decisive in that kind of war.

Eagle Flight was not decisive. But it stands as a good example on which to model some special counterambush tactics in brushfire wars throughout the world. Although it seems most suited to military operations in the countryside, Eagle Flight methodology has some applicability to counterterrorist operations in metropolitan and suburban areas. It can get a counterambush force on site faster and deliver more firepower to break up a guerrilla or urban terrorist attack than any other method yet devised.

It is designed to do one or all of the following:

1. Complement the operations of committed heliborne or ground forces
2. Extend the combat effectiveness of such forces
3. Operate independently, either alone, or reinforced, on a variety of missions

Although it often seems to be used defensively, e.g., to break up an ambush and scatter and pursue the attackers, Eagle Flight can constitute in itself an airborne ambush force. It does so by locating a moving enemy force and swooping down to surprise them in an airmobile attack.

Those who named the tactic Eagle Flight claimed that they did so because, as its name implied, it was a force capable of searching while soaring, pursuing its prey, attacking in force, and withdrawing to seek and attack more prey.

That's a poetic outlook not generally found among military minds. That description also tends to obscure the reality and the potency of the concept. An Eagle

Flight operation is a tactical concept involving the employment of a small, self-contained, highly trained helicopter-borne force.

The force can locate and engage the enemy or pursue and attack an enemy fleeing larger friendly forces. Ambush-busting involves both attack/engage and pursuit/attack techniques, making this an unusually useful method of breaking down ambushes and making them too costly for the enemy to continue. As an airmobile force, Eagle is also prepared to engage an enemy located or fixed by friendly forces. The ability to commit to battle immediately, either alone or in conjunction with other forces, is its most significant feature.

The generally flat terrain of the Delta region of South Vietnam was ideally suited to helicopter employment. With the advent of the helicopters now flying at night with night vision goggles, this tactic would be even more effective today.

Conversely, the lack of adequate roads, the great expanses of land that was inundated during the monsoon season, the vast network of tree-lined rivers and canals, and dense yet widely dispersed population made Vietnam a most difficult area of operations for conventionally equipped land forces. Not surprisingly, those same hindrances to operations by government and allied forces made the Delta region an ideal area for insurgents.

Finding, fixing, and destroying the enemy thus became monumental tasks—ones that challenged the imagination of tactical planners. While Eagle Flight operations were created for this environment and were tailored to the conditions in this particular area of operations, they are equally adaptable to any area suited to helicopter operations in which the location, pursuit, and destruction of enemy forces are a principal problem. Put another way, they are the ideal counterambush operation.

The commitment to Eagle Flight operations on an around-the-clock basis is considerable. The basic Eagle force consists of four squads of soldiers (or paramilitary police), plus command and advisory elements, i.e., company commander, executive officer, four squad leaders, artillery forward observers (AFOs). They are mounted in four troop-carrying helicopters. The Eagle force is normally supported by a flight of five armed escort helicopters.

The first helicopter carries the Eagle Leader company commander, radio operator, and AFO, as well as the first squad. (For identification purposes in expediting loading and as an aid to command control during operations, each person in this helicopter wears a *red* scarf or piece of cloth affixed to his uniform.) Aboard helicopter number two is a squad leader with a radio and the second squad. Each person on this aircraft wears a *green* scarf or piece of cloth. Helicopter number three carries the unit executive officer with a radio, a squad leader, and the third squad. Each person wears a *blue* scarf or piece of cloth. Number four carries a squad leader equipped with a radio. The squad members aboard the fourth helicopter wear a *yellow* piece of cloth or scarf for identification.

In practice, Eagle Flight carries out a continued reconnaissance in force and provides support to ground troops. An ambushed unit may call in an Eagle Flight force. Or, reinforced by fighter aircraft and supported by light fixed-wing observation aircraft, one of these units can be assigned the mission of probing for rebels/terrorists in several 25–50 square mile areas, depending on the population density in the areas.

Sectors of search are assigned to the observation aircraft. Operating "on the deck," the observation aircraft reports any fleeing groups, armed persons, camouflaged individuals and positions, concentrations of small craft, and the general reactions of persons in the search areas. It also recommends possible landing zones.

The Eagle Flight force commander selects the best targets while orbiting in the lead helicopter, performs closer inspection of the potential objectives, and coordinates landing and/or assault plans with the armed escort helicopters and fighter aircraft.

The armed helicopters normally make assault passes prior to landings by the troop-carrying helicopters. Then the Eagle Flight force lands; it contacts the enemy or screens the area and interrogates civilians. Reports are continually made to higher headquarters by the Eagle force commander, using the airborne observation aircraft for radio relay. The observation aircraft also scouts beyond the area, attempting to detect enemy forces who have reacted to the Eagle force landing, either by fleeing or preparing to attack.

If there are no contacts from a landing or upon completion of the action, a pickup is arranged. The troop-carrying helicopters, under the cover of the armed escort, complete the pickup. Eagle force is ready to continue its search or pursue any enemy reported by the observation aircraft or ambushed units. Remember, though, that pursuing the enemy leaves one's own unit vulnerable to ambushes set by the pursued. It was common for SEALs to break pursuit in Vietnam by quickly planting claymores with 10-second time fuzes on their trail to discourage or maim pursuers.

When the Eagle force is airborne, no more than five minutes' flying time is needed to reach the most distant sector of the ground operation. Within an additional three minutes or fewer, the landing can be planned, coordinated, and executed by the Eagle force and its armed escort. Fighter bombers can also join in the attack. That much firepower, when it can be concentrated in such a short period of time, discourages would-be ambushers.

In fact, the Eagle Flight force can be used with ground forces to set up an ambush of ambushers. A unit of infantry can serve as a Judas goat or stalking horse while the Eagle Flight can reinforce a ground force by being placed on-station to operate in areas adjacent to the committed unit. The armed escorts can perform low-level search and target-marking missions.

One of the main attractions of an Eagle Flight force is that it can even reinforce itself. When heavy opposition is encountered by a committed Eagle force or a requirement arises to block an exit in from the area of contact, the Eagle force can reinforce itself quickly by using its four empty troop-carrying helicopters to bring additional squads into action. Within a few minutes of receiving a request for assistance relayed from the Eagle force, four squads of soldiers from the nearest unengaged friendly unit can be ready for pickup. Since the helicopters will be in radio contact with the requesting force's commander, the reinforcements can be briefed on the situation, assigned objectives, and given landing instructions while en route.

Eagle forces are not invincible. When properly planned and structured, they will have shock power from their five gunship consorts, the maneuverability that comes with being an airborne force, and the stability of infantry operations. But they will need backing up. And Eagle forces do have vulnerabilities.

After landing, the force's combat power is limited by the skill and firepower of the squads and the support to be expected from the helicopter gunships. The force itself usually carries no heavy machine guns or mortars. To prevent being ambushed or defeated in detail, a massed landing should be sought wherever possible.

Often the force is separated by a terrain feature, e.g., canal or tree line. If upon landing these small forces are surprised by a large enemy force, they cannot be extricated easily because the helicopters would be extremely vulnerable to loss if called in during a heavy firefight. Unless the commander is reasonably certain of the size, strength, and disposition of the enemy force, a landing zone should be selected that will place the force beyond effective small-arms range of the enemy.

The Eagle force cannot move by stealth because its mode of travel advertises its presence well in advance. Few things in war, short of an artillery barrage or an on-the-deck overflight by supersonic aircraft, make as much noise as nine whirlybirds belching and chattering their way over the countryside.

The force is dependent upon considerable support; the more independent its mission the more support it requires. Coordination of the force and its supporting elements is of paramount importance.

Water features complicate the tactical picture for Eagle Flight forces. In attacking an enemy force located near a canal, a landing formation should be selected that will place part of the Eagle force on the opposite bank to block escape. The Eagle force should not attempt to unload and assault in flooded areas where the water is more than 3-feet deep. When troops have to go into flooded areas, they should be landed on adjacent high ground and commandeer boats or sampans for use to complete their mission.

Special techniques of planning and employment must be used in Eagle force operations. Experience has shown that several factors must be considered in planning for a landing. The planning must be accomplished immediately upon arrival over the selected target area. It must be thorough but done quickly, because the enemy will already have started to react to the presence of the attacking force by the time the helicopters arrive on the scene. Thus, planning, reaching decisions, and communicating these decisions in the form of instructions to the force and its supporting elements must be virtually simultaneous.

The following factors always need to be considered by the Eagle force commander:

1. Are the suspected personnel actually enemy forces? In Vietnam, for instance, there was always a great danger of mistaking friendly Self-Defense Corps

(SDC) personnel—armed and dressed in black or nondescript uniforms—for VC. Generally the VC would run, attempt to hide, or fire at the helicopters; SDC or Civil Guard soldiers would generally wave for recognition. Lack of communication methods with the ground and a positive means of identification compound the identification problem.

2. What is the proximity of the target to heavy vegetation? How many enemy are visible? In selecting the landing zone, care must be used to avoid landing in a preplanned ambush, one based around hidden enemy forces.

3. Is the enemy force massed or scattered, organized to fight or disorganized? Decision regarding use of a preassault airstrike, choice of landing formation, selection of landing zone(s), and decisions regarding the probability of success when using a split force against the enemy are affected.

4. Is the logical target bisected by a canal or similar terrain feature? Again, the choice of landing formation, selection of landing zone(s), and decision regarding capabilities of split forces against the enemy are affected.

5. If the area is inundated, as in a swamp or paddies, is the water too deep for the force to maneuver effectively? Selection of landing zone is affected.

Landing formations must be decided upon. There is a wide range of choices of landing formations. Selections must be based on nature and size of target and the terrain features in the target area. Experience in South Vietnam operations showed that the following choices against specific targets were workable:

- To assault enemy forces in groups of up to 50, disorganized and in the open, depending upon how heavily armed they are, a landing formation known as the "half box" is chosen. It provides a "three o'clock exit" for all troops, thus eliminating the necessity for any troops to move around the helicopters before assaulting. The enemy is caught in the crossfire between the two lines of assaulting forces.

• To screen a large open area following an airstrike, or to search for VC hiding underwater or in high grass, the Eagle force troop-carrying helicopters are landed in "line" formation, with about 100 yards between helicopters.

• To attack a large enemy force reported to be in dense vegetation, the helicopters are landed in line formation about 300 yards away. A closer landing has the tendency to place the formation in an enemy ambush. If the enemy force is reported to be small, the helicopters are landed in line much closer to prevent enemy from escaping before assault.

• To trap an enemy force hiding in groves along the banks of a canal, the "open box" landing formation is used. Two aircraft land on each side of the canal forming a box roughly 300 yards on a side. Squads assault and direct their fire at the enclosed target.

After the commander selects his target and decides on his landing formation and plan of assault, he has to communicate his decisions to the helicopter pilots and the armed escort helicopter flight leader. One key problem that must be overcome by the force commander seems simple but is vitally important. How do you describe what the target is? Helicopter pilots may have problems understanding what the commander of the Eagle force wants.

One obvious means of briefing is the oral description. The use of compass and clock directions in verbally describing a target and landing zone is useful—it is mandatory in an area dense with canals and groves since the mere reference to a canal would be meaningless. A typical landing instruction to helicopter leader might be: "The canal running from NE to SW about 500 meters out on your 2 o'clock position is the baseline. Do you see it? At 3 o'clock on the baseline, a small clump of trees on the canal is the center of mass. Do you see it? Land numbers one and two on the NW side and numbers three and four on the SE. Keep both sections 300 meters out from the canal."

Another means of communicating intentions is by marking with tracer ammunition or smoke grenades. Helicopters can also be vectored over target by pilots of other aircraft.

Communicating with the pilots is one problem. Communicating with the troops is another. A unit loaded onto four helicopters, each of which is making enough racket to deafen everyone within hundreds of yards, cannot be properly briefed. Since the forces have not been thoroughly briefed on either the landing formation or the assault plan, unless ground winds absolutely prohibit the landing direction, pilots always attempt to land in a direction that provides the troops a "3 o'clock" exit toward the objective. If they cannot make such a landing, crew chiefs draw a directional arrow on a piece of paper indicating the direction of the objective as it will appear to the soldiers when they emerge from the helicopter.

The last-minute orientation on the direction of the objective, prior to unloading at the target area, can also be carried out by the use of simple hand and whistle signals.

Immediately upon landing, squads are assembled rapidly, and control must be quickly gained by the commander and his squad leaders. As this is taking place on the ground, a designated officer immediately establishes radio contact with the orbiting empty helicopters and, through the O-1 type observation aircraft radio relay, reports to the division headquarters.

On the objective, the troops must work rapidly to screen the area, kill or capture the enemy, apprehend suspects, and prepare to reload. Since an Eagle force is more effective when airborne and prepared to attack, excessive time is not spent on the ground following the assault or capture of enemy. If a more lucrative target is reported to the empty orbiting helicopters by other units, a red smoke signal is dropped as a signal for pickup.

Upon completion of the ground action, the commander studies the area and determines the center of mass of each squad. If terrain permits, the troops are formed in a column of squads, approximately 30 yards between squads, with the first squad (red) up wind. However if a squad is widely separated from the force, possibly separated by a canal, it can be reloaded in place providing a suitable landing place is available. The designated officer contacts the helicopter flight leader by radio, informs him of the wind direction, and identifies the landing zone(s). It is SOP that, should radio communication between ground and helicopters fail, forming up troops in line of squads in pickup formation is the signal that the area is secure and force is ready to reload. As described in the loading plan, each squad has a color code that is indicated by a scarf or piece of cloth worn on their uniforms (red—first; green—second; blue—third; yellow—fourth). Each helicopter is designated by a corresponding color code. Thus, identification for pickup and relocation is simplified.

While awaiting return of the helicopters and during all loading operations, each squad maintains close security of the area. The armed escort helicopters continue to orbit, protecting the helicopters and troops during reloading operations. Reloading is nearly always done while the helicopters maintain partial power.

If the helicopter is carrying a nearly full fuel load (early in the operation) or if prisoners aboard cause the load to exceed the normal number of men, an obstacle-clear distance of hundreds of meters in an upwind direction will be required for takeoff. Each helicopter should take off as soon as loaded in order to minimize time on the ground, a period of great vulnerability.

NOTE: Prior to Eagle Flight operations, all seats are removed from the cargo compartments of the troop-carrying helicopters. During the operations, no equipment is tied down.

HITTING BACK

The key to successful counterambushes is striking back. Pursuit must be initiated with the least possible delay, with only that degree of caution required to prevent falling into a larger, and perhaps the primary, ambush.

Military studies show that relief and pursuit, executed automatically as a matter of first priority, is most important in the overall effort to reduce the effectiveness and frequency of ambushes. First, striking back ensures an early relief of the ambushed unit. Second, it increases the possibility of friendly forces making contact with the ambush party before it disperses. Third, it reduces the time available to the attackers to destroy the ambushed force and to loot the vehicles. Finally, successful pursuit operations will improve the morale of the friendly units while having an opposite effect upon the ambushers.

The counterattack was always the weak point in the whole U.S. strategy in Vietnam. It is probably the weakest part of any military counterambush program. And counterattack is certainly not an option in most executive ambush situations, whether the protection program is designed to cover a visiting dignitary in Guatemala against leftist rebels or a banker in Bonn. Even if a bodyguard response force had both the manpower and mobility to take off after a group of terrorists who had attacked an executive under their protection, governments get justifiably upset at the prospect of people chasing people through the streets firing automatic weapons at each other. "Hitting back" in executive ambushes is a matter best left to police.

In military operations, a helicopter or parachute alert force, positioned with aircraft ready for instant employment, provides an excellent relief and pursuit capability. Eagle forces are but one type of airborne quick-response unit. A request for tactical air support should be included automatically in the alert message unless aircraft are providing column cover or otherwise are immediately available at the time the ambush is sprung. Armored carriers and armored reconnaissance vehicles may also be used to advantage in quick-pursuit situations since they, too, have a psychological effect. Although ground movement by foot or truck should not be overlooked, organizing and getting a convoy or column to an ambush site generally takes too long for the newcomers to effectively pursue, pin, and destroy the ambushers.

In the case of an executive ambush, there will normally be no pursuit or counterattack phase, except that mounted by police forces.

IMMEDIATE-ACTION PROCEDURES

Every ambush is unique. There is no textbook response that will cover every situation. Keeping in mind that elsewhere in this book some other alternatives are recommended for roughly similar situations, there is a fairly standard Immediate-Action Procedure (IAP) for military ambushes of convoys.

The procedure is a drill, one that should so ingrain a pattern of action and thinking in troops that they can perform without orders. Of course when special situations require it, the actual response can be modified through real-time orders. The fact is that no matter what precautions are taken and no matter what preparations are made, the properly planned ambush will *always* be an unexpected encounter. Troops need to be grounded in the baseline response that the immediate-action drill gives them.

The immediate-action drill outlined here consists of

simple courses of action to deal with the problem of the unexpected encounter. The aim of these moves is to neutralize the ambush and regain the initiative through immediate and positive action. The drill is based on the experience that it is a fatal or near-fatal mistake to halt in an area that ambushers have carefully selected to be the killing zone.

The basic IAP is to keep moving when fired on, to halt only after getting through the killing zone or before entering it, and to counterattack immediately from the flanks chosen by ambush victims.

1. Drivers should not stop, but should attempt to reach positions that are clear of fire.

2. Lookouts and guards should immediately fire on the known or suspected ambush positions. Troops in each vehicle will fire all available weapons to disrupt and confuse the enemy.

3. As vehicles clear the killing zone, they should stop to permit occupants to detruck in order to take immediate offensive action.

4. Vehicles other than the armored escort should not attempt to run the gauntlet of the ambush. Commanders halted outside the killing zone should detruck their troops in order to take immediate offensive action.

5. Elements of the convoy that are under attack will report their situation immediately to other elements of the convoy or other friendly forces that may provide support.

6. Troops in vehicles unable to drive clear of the killing zone should immediately open fire and launch grenades at the estimated ambush positions. They should dismount from the vehicle on command and attack ambush positions. Troops in "hardened" vehicles may not be required to evacuate the vehicle immediately. They will fire all available weapons at guerrilla ambush positions and wait for the first lull in the initial guerrilla fire or for supporting fire from friendly armored escorts

or area-fire type weapons. Disembarking the truck should then be done in the same manner as from a soft vehicle.

IAPs should be rehearsed frequently by infantry units. When miscellaneous vehicles are formed into a convoy, two or three drills should be staged before the convoy enters a danger zone.

LEBANESE CAR BOMB AMBUSHES

This type of ambush is used in both military campaigns and in executive ambushes mounted by terrorists. The roadside bomb ambush is virtually the same, except that no car is involved.

In the Lebanese car bomb attack, a car, truck, cycle, or van packed with explosives is left at some location. When a car or convoy carrying the ambush target rolls past, the explosives are set off by remote control.

Other variations of this tactic involve timed ignition on a city street or in a shopping center parking lot to create random terror. The ambush "target" is whoever happens to be in the location at the time.

Yet another variation uses a suicide bomber. The suicide bomber variation has been used with good effect by Shiite Muslim groups against Israeli military convoys; in its "raid" form, it was used with deadly effect against U.S. and French targets in Lebanon. The Liberation Tigers of Tamil Eelam also use the suicide car bomb/cycle bomb with deadly effect. (See also Roadside Car Bomb Ambush, Executive Ambush, Armored Vehicles.)

LEAVING THE HOUSE

The dangers of life are magnified by the need to move. At home or in the office, the potential terrorist target is protected by locks and walls, at the very least. Attackers have to mount a raid to get at the target. There is also an invisible wall of social norms dealing with activities in private, one that often provides a warning at the office or home that something is amiss.

Leaving the house by the front door, walking out of the protective cocoon, leaves the individual as vulnerable as a butterfly in a windstorm. The more predictable the target's movement, the greater chance there is for an ambusher to capitalize on it. There are few things more predictable than:

- Getting into a car and driving to work
- Leaving the office for lunch at a favorite restaurant
- Driving home at the end of the day

These generally vulnerable movements can be divided into two categories: those made on foot and those made in a vehicle. Either can be exploited by ambushers. Narco-terrorists in Colombia gunned down a judge while she got out to open the garage door when returning from work (a garage door opener has its advantages). A top African National Congress leader who had a background in mounting ambushes was shot down between his car and his front door by a right-winger in South Africa. A former prime minister of India and one of Sri Lanka's top leaders died while on their feet at political demonstrations, blown away by suicide bomb ambushes mounted by the Liberation Tigers of Tamil Eelam.

Successful ambushes against people in cars are almost too numerous to mention, from German bankers to U.S. military officers serving in foreign duty stations.

The key to survival in an ambush, particularly for individuals targeted in executive ambushes, does not lie in breaking up an ambush once it has started. If there are keys, one is to be relatively invulnerable. Making oneself difficult to ambush forces the terrorists to attack someone else. The second key is to identify an impending attack and the attackers while they are still in the planning stages.

It is a truism that virtually all successful terrorist

ambushes involve extensive surveillance of the target, identification of his or her most vulnerable movements, and development of an attack plan that capitalizes on those vulnerabilities.

The self-protection program designed to frustrate executive ambushes has to be aimed at both the identification of possible surveillance and the use of protective behaviors.

It's important to be logical in planning, but also important to be flexible in carrying out a self-protection program. Possible targets should never overrule gut feelings about their safety. Whenever intuition says something is not quite right, even when the exact cause of that feeling can't be pinpointed, follow the feeling. In the case of leaving the house, that means that if an executive has concerns about safety on a particular morning, it's a good day to change plans for the day. It may be a good day not to even go to the office.

As a matter of routine—whether leaving home, the office, or some other spot that is visited frequently—a potential target or protected person should check the surrounding area carefully.

It's easiest to do at home. The potential ambush target observes, not just looks at, the surrounding area several times before leaving. He uses the doorway peephole, observes through angled blinds. He looks for people outside, makes note of any loiterers, door-to-door salespeople, street vendors, repair crews, and moving vans and crews. He looks for people sitting in cars or driving by repeatedly—particularly if they are slow to pass. He is particularly aware of people under 40; few middle-aged or geriatric terrorists do operational work or surveillance. Anything out of the ordinary should be noted in a logbook—dates, times, descriptions, and license numbers are useful items of information.

At home, the visual checks can be made on first getting up and every few minutes thereafter. But the checks should be concentrated, several of them should be made, in the final 10 minutes before leaving.

At the office—and particularly at other places like restaurants—it's going to be more difficult to make several comprehensive checks of this type before leaving. But it still should be done, within the obvious constraints and despite the difficulties.

An ambush target who sees anything during these checks that excites suspicion should simply not leave. The legitimacy of repair crews can be checked through the utility or government agency. Neighbors or nearby friends can check out salespeople going door-to-door or street vendors. A call to company security staff or police is worthwhile to deal with loiterers or people in parked cars.

After observing carefully from the inside and feeling secure, the executive should step outside and look around. This is particularly true for those living in an apartment building or working in an office building where fields of observation will be more limited than at a house. Looking up and down hallways for people who can't be identified, particularly new neighbors, is key. Listening for sounds, such as the rustle of clothing when no one can be seen, is important. Particular caution is needed in hallways where there are blind spots or recessed entryways.

The executive should sometimes make one or more false starts at leaving, then return and observe. Anyone watching may be drawn into action by a false start. They'll come out of hiding, start their car, or react in some other way to the movement.

For those who have a single-family home with a garage, the car should be kept inside the garage. The garage door should be kept closed and locked until it is time to take the car out. All garage doors should be equipped with automatic garage door openers so that no one has to stop, and get out to manually open and close the door and thereby make a clear target of himself.

Those who don't live in a single-family home with an attached garage should generally be parked as close as possible to the house—consistent with other safety considerations. For instance, a car should never be parked in a shrubbery-obscured parking bay, no matter how close that bay is to the door.

When going to the car, the executive should always use the safest path possible. That is not always the easiest path or the most direct route. If a car is some distance from the door, the path to the car should be varied from day to day. If anything suspicious is observed during that walk, the intelligent executive doesn't forge blindly ahead. Going back is one option; heading for the nearest place where there are people (a neighbor's home or a store) is another. From there, security personnel can be called for help or advice. In some cases, it's best to use some form of transportation other than the car. Calling a friend for a ride, using a cab, or taking a bus may be options.

When going to or from the car, the key to the next lock should always be out and ready for use. Fumbling to unlock the door leaves the executive exposed to ambush for long seconds. Always remember which way the lock turns (clockwise or counterclockwise) so that the motion is minimal to unlock the doors. Once in the car, the executive should lock the door, put the seat belt on, and prepare to move the vehicle immediately if anything suspicious occurs.

Tails—surveillance teams—can often be spotted by carefully observing anyone pulling out into traffic from the same general area. They will generally make their move within a block or two. An experienced tail may wait longer—if he can see the target—but the target should be able to see the tail as well.

On arriving at the destination, the executive checks for the same suspicious signs as before: loiterers, work crews, and so forth. Where possible, it is wise for an executive to call ahead by *secure car phone* or radio and ask someone there to check for suspicious activity. This is particularly helpful when returning home at nightfall or after dark.

The potential target should park as close to the place he intends to go as possible—again keeping other security principles in mind. For instance, it is unwise to leave cars in underground lots or in parking structures where there is little or no security. Any time that another vehicle can be parked next to an executive's car without the possible target knowing who owns it, there is a definite danger of an ambush mounted from a vehicle. Vans should be of particular concern. Where parking is not secure and the risk is high, changes should be made in the living or working arrangements to overcome the problems.

When leaving a car, the smart executive takes the safest—again not necessarily the most direct—route to his destination.

When entering a parking garage on foot and returning to the car, it is important to look over the area and check carefully for strange people, unusual activity, or suspicious sounds. If things look and feel suspicious or are not quite right, it is wisest to return to the starting place and call security officers or police, take alternative transportation, or try again later to see if things still look or feel suspicious.

In multistoried buildings, smart targets use the elevator when possible and the central stairwell where necessary. Stairways are inherently dangerous. The repeated turns create blind spots. Door and hallway exits off stairs make them excellent killing zones. When a stairway must be used, the one that has the most traffic on it is generally preferable; the traffic will tend to discourage loiterers.

Elevators require special care. Physical proximity that would be a definite, if late, danger signal on the street is the norm on an elevator. It goes without saying that the general rule is to never get on an elevator with anyone suspicious—and they do not have to be holding an Uzi to arouse suspicions. A stranger holding an elevator for someone counts as suspicious. The smart executive takes the next one.

Smart executives always try to punch the floor button themselves. If they can't do it because there is an elevator operator or because the elevator is too crowded to reach over to the button, they should watch to see that it was, in fact, activated. It's important to be aware of the location of the emergency button. And it's a wise idea for a potential target to get as close as possible to the emergency signal while being prepared to lunge for it if necessary.

When entering or leaving any building with a lobby, it is a good idea to stop and observe for a moment what people there are doing. It's also important that any potential ambush target watch what people do after he walks in, seeing if they follow.

Abnormal activities on the road or in buildings should be jotted down in a notebook, as they are at home.

In traveling to or from work, it is often wise to vary the method of travel—using different cars, cabs, and even informal car pools with friends. Buses should not be used with any regularity. Cabs, like elevators, require special treatment if used regularly. To avoid the possibility of one being placed there specifically to carry off a kidnapping, no one should take a cab that appears as if it is waiting for him, or one that pulls out from a parked position just down the street. Likewise, it is dangerous to wait for a cab in the open or to try to hail one. The safest method is to phone a well-known cab company and wait inside the building for the cab to drive up.

LOCATING POSITIONS WITH THE HELP OF AIRCRAFT

In using fixed-wing aircraft to locate their positions on the ground, units sometimes employ colored smoke to mark their position. This not only identifies their position to the pilot, but also to the enemy, and makes it easier to set up an ambush.

Colored smoke should not be used to establish ground locations for pilot identification unless contact with the enemy has been made. In lieu of colored smoke, colored panels are available and should be carried for this purpose. Another method that can be employed when there is a thick jungle canopy, as there was in Vietnam, is to have the aircraft fly a crisscross pattern over the general area and to signal the plane by radio when it is directly overhead.

MARKING TARGETS FOR AIRCRAFT

Aircraft are a great comfort in countering an ambush—until the "friendly" aircraft start dumping ordinance on you. Friendly-fire casualties are the saddest kind. And in more than a few cases armed helicopters and fixed-wing aircraft firing suppressive fires have killed or wounded friendly personnel. This often resulted from inaccurate target designation or poorly defined locations of friendly troops.

Coordination between ground and air requires that certain signals be predetermined. One of the most important is to clearly mark friendly troop locations. The best marking means include panels or, in the case of friendly/enemy engagement, colored smoke grenades and rifle grenade streamers.

MEDEVAC OPERATIONS

Casualties aren't something anyone wants to talk about, but they are a grim part of ambushes on both sides. Uncleared weapons can discharge during medevac operations, causing casualties and damage. Weapons of dead and wounded should be cleared prior to loading on evacuation vehicles or helicopters. A Browning Automatic Rifle discharging during a medical evacuation can easily increase the casualty count, and on occasions they have.

MILITARY OPERATIONS IN URBANIZED TERRAIN (MOUT)

Warfare in urban terrain involves a great deal of ambush activity. This field, known by the acronym MOUT, is an entire specialty in warfare. The temptation is strong, for that reason, to either write an entire book on the subject or brush lightly over it with only one sentence. In fact, discussion of the ambush in urban warfare deserves more than a mention though less than a book, or even a chapter.

War in the urban environment is characterized by street fighting, sniping, and ambushes of small units and individuals as they move. Actually, the term ambush doesn't quite explain the situation because soldiers fighting on the urban battlefield expect to be attacked at any and all times. But the front is sometimes fluid and much of the front-line area is a no-man's-land.

Defenders, as a general rule, have increased opportunities of ambush and have the element of surprise on their side.

Buildings above and underground passages below ground level (from sewers to cellars) give warfare in the urban environment a vertical dimension.

The urban terrain in wartime is far different than deserts, forests, or marshes. The terrain varies greatly, even within a small town. There will be lighter construction in new areas; older areas will have more heavy construction. That will affect ambush positions. Rubble creates obstacles; that tends to strengthen the defender and impede the attacker. Since the roof is the weakest part of the building structurally, the top floor is more vulnerable than the lower stories. For that reason, the ground floor is the safest floor for soldiers to occupy. Nonetheless, snipers and observers can occupy high points.

Units fighting in urban terrain will find that command, control, and communications difficulties will throw greater weight on junior leaders—officers and noncommissioned officers (NCOs). This is small-unit warfare and shares much in common with ambushes. Communications with headquarters above company level are unreliable. Local commanders will have to communicate to their troops as best they can. Field telephones and messengers will be the most used methods of communication. Voice commands get lost in the noise of battles inside and around buildings. Radio communications is degraded by high concentrations of metal and a lack of line of sight. Fire and smoke from burning buildings makes pyrotechnic and smoke signals hard to see and interpret.

In the urban environment, vehicles are more vulnerable to short-range attacks—under 100 yards in most cases—but targets are exposed for briefer periods.

Automatic weapons, grenades, hand-placed explosives, and rocket launchers are most effective. But special care must be taken with weapons. Since rifle rounds will go through most interior walls and many exterior walls without trouble, special care must be taken that missed rounds don't cause friendly casualties nearby. Even hand grenades pose problems in room-by-room fighting. It's easy to think that just lobbing one inside will clear out any bushwhackers. Many grenades are too powerful for indoor use; the shrapnel will come through the walls and closed doors and injure anyone nearby. Stun grenades can be very effective in these settings. Grazing fire is difficult to obtain because piles of rubble, stubs of walls, wreckage, and fences form obstacles. Around tall buildings, mortars are often more effective than artillery because of their high trajectory. But mortar rounds have little effect on solidly built structures. They are most effective where the buildings are lightly constructed and in open areas. Weapons with back blast require special placement, and weapons whose projectiles arm themselves only after traveling some set distance can't be used against targets that are too close.

Underground passages—tunnels, sewers, etc.—are often used as supply routes and for cache sites. For that reason, they are prime ambush locations as well.

MINESWEEPING IN WATERBORNE OPERATIONS

The VC enjoyed considerable success in ambushing craft with electrically detonated mines laid in rivers and canals. The mine itself was sometimes the only ambush (a demolition ambush); that is, the ambush position was not always covered by a fire team. VC use of electrically detonated mines in rivers and canals often occurred when craft were returning from operations in areas where there was only one waterway available.

Minesweeping operations should be standard procedure during water travel in unsecured areas and when troops are embarked. When it is necessary to use the same route to travel to and from an operation, the route should be patrolled during the action to guard against mine emplacement. Minesweeping operations and patrols to prevent the introduction of mines into waterways are essential where only one waterway is available.

Despite all the sound advice to the contrary, the same route must often be used for transit to and from an operational area, or for patrol. Minesweeping is particularly important in these cases. (If the route is short, if boats abound, and if the operation or move is of short duration, stay-behind boats strategically placed can help keep the return route secure.)

Mines used in boat ambushes are generally either command detonated or of the contact type. Detection of mines before they can cause damage to a patrol craft and other vessels on a river or estuary is essential.

Positions of underwater mines are sometimes marked by small floats made of wood or Styrofoam, very similar to the normal fishing floats seen in the river. For that reason, it is important to be aware of fishing floats and consider them to be markers for mines.

It is also important to be aware of any people, particularly partially hidden or camouflaged people, as they may be responsible for setting off the explosives. They may be nearby, but they may well be some distance away. Detonation wires may run along the river bottom for a distance of several hundred yards. Thus,when a mine explodes, the entire riverbank for a distance of several hundred yards in both directions should be surveyed for indications of the withdrawal of the person detonating the mine. Firing on the area immediately adjacent to the location of the explosion, which is the normal procedure, sometimes may not be the best retaliation.

MOTORCYCLE AMBUSHES

Many modern terrorist ambushes are mounted from motorcycles. The attacker rides as a passenger on a cycle, which usually approaches the target from the rear at high speed. The attacking passenger sprays the target vehicle with automatic weapons fire. Both cyclist and attacker then roar away before anyone has an opportunity to react. This type of ambush can generally be defeated if the cyclist is seen in time. If the driver of the target vehicle (perhaps warned by the sound of the rapidly approaching cycle) senses the attack developing, a simple flip of the steering wheel can defeat it. Turning quickly but moderately in the direction of the cyclist will cut him off and force him to take evasive action. That will almost inevitably throw off the aim of the gunman.

NIGHT AMBUSHES

The principles of daylight ambush operations apply to night ambushes as well. However, at night some modifications may be required. Concealment is plentiful at night, but observation is limited and fire is less accurate. Therefore, weapons must be properly sited to ensure complete coverage of the killing zone with fire. Weapon fields of fire should be fixed by stakes. Positions should be closer together for better control. Ambush positions should be occupied at night but after a daylight reconnaissance if feasible. Flares should be used, when available, to support the ambush force. Infrared imaging and starlight scope weapon sights may be used by ambush forces to view personnel and objects in the dark. They make it possible to fire at appropriate targets in darkness and to send and receive predetermined code signals by using the light source to transmit and the telescope to receive. If required and available, luminous tape or paint markings may be used for identification.

NIGHT DEFENSIVE POSITIONS AND AMBUSHES

Aside from local security, ambush positions should be located no less than 1,500 to 3,000 feet from night defensive positions.

PATROL MOVEMENT TO AVOID AMBUSHES

Patrols are much different than convoys. Being in a patrol means that slightly different counterambush and anti-ambush procedures need to be used. For instance, movements on trails should be avoided where possible. Trails provide the initial channeling that leads ultimately into a killing zone. It is often necessary to move on trails, particularly in mountainous country, but there are things that can be done to reduce the danger. Whenever a halt is called along a trail or track, it is important to move well clear of it and adopt an anti-ambush position.

The mission of a patrol also requires some modification in the way potential ambushes are treated. Although traditional counterambush procedure requires that the first person to spy an enemy opens fire immediately, while indicating the direction of the enemy, the mission of the patrol calls for a different approach. The enemy may not be an ambusher—and may not even be aware the patrol is there. Carrying out an immediate attack reveals the presence of the patrol and compromises the mission. If the patrol spots an enemy, its troops should freeze like an animal does, then slowly ease out of sight. Sudden movements can catch the enemy's eye and pinpoint that patrol's position.

When stopping for any reason—long or short halts, when setting up an ambush—a patrol should always establish an all-around defense. There are no exceptions to this rule.

In entering a danger zone—roads, open areas, fences, river or stream crossings—extra security precautions are necessary to avoid ambush. Typically, the point man, the scout preceding the unit, will identify a danger area to the patrol leader, who will then usually halt the patrol and call up the designated security element (often an automatic weapons man or grenadier or combination of people filling those slots).

The patrol often moves into a box formation when passing through danger zones. It is the best formation to cross or fight. The patrol will resume a single-file formation when the danger is behind.

There are a number of ways to cross a trail or road, depending on the situation:

- The element or unit forms a skirmish line and advances across the road at a fast walk.
- The members of the element form a file and cross quickly and quietly.
- Patrol members cross the road a few at a time until the entire unit has crossed.

Crossing streams is similar to crossing roads, but all equipment must be secured and waterproofed prior to going out on a patrol where stream operations are possible.

POST-AMBUSH AMBUSHES

In many cases an ambush does not lure all of the enemy into the killing zone. As a result, part of the enemy element will escape. Often the ambushed group will return to the site of the ambush to retrieve weapons and bodies. This presents a good opportunity to spring a second ambush. As soon as a unit springs an ambush, a team or a larger unit should move in the direction of the enemy withdrawal. About 700 feet from the site of the first ambush, a second ambush should be hastily prepared. In Vietnam, this was one of the more effective methods of attack against the VC, who almost inevitably returned.

PREPARATION OF CONVOY VEHICLES

Armed troops in a military vehicle should have all-around visibility. They should be able to fire their weapons without hindrance and detruck quickly. Where possible, armored vehicles should be used. Often there will be no armored vehicles available. "Soft" vehicles should be "hardened" as much as possible using things such as sandbags and armor plating. One simple and easy method of hardening a soft vehicle such as a 2 1/2-ton truck is to place a single row of sandbags, stacked five deep, down each side of the truck. That barrier will stop most small arms projectiles. A wooden bench rigged down the middle of the truck bed allows troops to sit facing outward.

Other preparations of convoy vehicles include:

1. Removing tarpaulins and bows.
2. Removing the tailgate or rigging it to drop to the horizontal position instantly.

3. Placing chicken wire over the open windows of larger trucks so that ambushers can't toss grenades into the cabs.
4. Attaching a cutting or deflecting bar at the front of vehicles to prevent barrier cables or wires stretched across the road from decapitating or otherwise injuring those riding in the vehicle. This is particularly important in the case of smaller vehicles, such as the 1/4-ton truck.
5. Protecting beds of trucks and floors of vehicle cabs by at least one layer of sandbags to minimize casualties from any mines detonating under the vehicle.

PRISONER CONTROL

When prisoners are to be seized, one or more members of the ambush team is designated as prison handler, or PH. Proper prisoner-handling equipment, such as handcuffs or ties, should be brought along. Blindfolds and gags may be needed as well. In the field, the work of a PH is trickier than handling hydrochloric acid. The job is characterized by the "Five S's."

1. Seize
2. Search
3. Secure
4. Silence
5. Segregate

When seizing a prisoner, it is important not to injure him to the point that he can't walk or keep up the pace of the rest of the ambush team on withdrawal. If he is injured badly he will have to be carried.

The prisoner should be searched immediately, and all papers, web gear, and potential weapons such as shoelaces should be taken away. Prisoners should always be searched by one person while another provides security with a loaded weapon pointed at the prisoner. The person conducting the search should never get in the way of the security person.

When possible, keep the prisoner spread-eagled on the ground and search thoroughly. Always walk around the feet; a prisoner can reach out and grab or trip a searcher who is near his head. The spread-eagled position also allows the searcher to stomp on the genitalia if force has to be used to maintain control.

Watching the prisoner's eyes during the search will often alert a searcher to any escape attempts, hidden weapons, or documents. People have a tendency to glance in directions where they have concerns—whatever those concerns are.

Maintain control by speech or sign language—and by force if necessary. The prisoner should not be allowed to talk, move freely, look back, or distract any member of the force.

RAID-AMBUSH PATTERN

An ambush is executed many times in conjunction with the attack of an isolated outpost. The pattern is to attack the outpost and at the same time establish an ambush site along the route(s) that the relieving forces will most likely follow.

RAILROAD AMBUSHES

This is a special kind of ambush that includes the elements of sabotage. Lawrence of Arabia made his mark here; the tactics haven't improved or evolved significantly since the days of his exploits.

Lines of communication are difficult to secure in areas where guerrillas operate. Limited manpower usually prevents adequate security for long lines of communication, such as railroads.

A key point in railway attacks is remembering that the

rails themselves constitute the vulnerable choke point. The movement of a train is directly determined by the condition of the rails.

From the ambusher's point of view, moving trains may be subjected to harassing fire, but the most effective ambush involves derailing the train. The locomotive should be derailed on a down grade, at a sharp curve, or on a high bridge. This causes the cars to overturn and results in extensive casualties among passengers and train crew members. In the case of freight trains, the goods inside the cars are more likely to be damaged, tanks are more likely to rupture, and so forth.

When attacking passenger trains, fire should be directed at the exits of overturned coaches while designated groups armed with automatic weapons and grenades rush those coaches that remain standing—whether they are on or off the track. Other groups salvage usable supplies from freight cars and torch the train.

After the ambush, rails are removed from the track at some distance from the ambush site in each direction to delay arrival of reinforcements by train. If the line is double-tracked, rails need to be removed from each set of tracks.

Since trains, unlike trucks and cars, cannot deviate even an inch from their assigned track, they have no flexibility. They either go forward or backward along the track. Where insurgent forces are large enough and have good enough communications to stage multiple attacks, that inability of the train to move any way except forward or back offers ambushers some unique opportunities. Once a train is moving along a stretch of rail, attackers can isolate the line. Destruction of track, ambushes, etc., on either end traps the train. Often it is difficult to get reinforcements to the train, except by rail. Having isolated the train, insurgents can either stage a raid against the train—if it stops—or stage an ambush at some place along the line if it keeps moving back and forth.

When planning the ambush of a train, remember that

the enemy may include armored railroad cars in the train for its protection. Important trains may be preceded by advance guard locomotives or inspection cars to check the track. In some cases, parties on foot may check culverts and trestles—the most likely places—for mines and explosives.

When viewed through defensive eyes—from the security leader's or military commander's point of view—the situation becomes much more convoluted. Railways are a major national resource. Anything that makes them unusable or threatens to keep freight and passengers from moving over them is a serious threat to the nation's economy and the government.

If railways come under ambush attacks, measures will have to be taken to protect the rail line, designated installations, railway repair and maintenance crews, and the rail traffic itself.

Three of those four categories involve counterambush planning, and designated installations (such as bridges, underpasses, tunnels, towers, yards, and roundhouses) have to be protected against raids. Where possible, to protect against raids, sabotage, and ambushes along critical lines, the right-of-way and an area about 300 yards—or the effective range of small arms—on either side of the tracks is declared a restricted zone by the authorities. After it is posted, civilians living within the zone are evacuated and unauthorized persons are forbidden to enter.

Where clearance of the entire railroad right-of-way is impractical—and that is most often the case—areas around critical locations such as defiles, tunnels, and bridges may be cleared. In critical places particularly, underbrush and thick forests must be removed from the sides of the roadbed.

The most effective counterambush tactic—and that is saying very little at that—is to establish block houses and security units along the right-of-way. Frequent patrols must be made along the right of way and the flanks to discourage trespassing, give warning of an impending guerrilla operation, and detect mines and sabotage along the rail bed. The latter may be a sign of an impending ambush as well as simple sabotage.

Armored railroad cars may be used to supplement such patrols. Railway inspection, maintenance, and repair crews have to be provided with armed security detachments to fend off ambushes. Those crews constitute a choke point on a railway operation and need to be cared for with all the concern shown a major infrastructure target, such as a bridge or tunnel.

When faced with a pattern of railway ambushes, authorities should switch to running trains on an irregular schedule.

Trains should also be provided with security detachments to defend them against ambushes. Railroad security elements should precede and follow individual trains. Train guards may be assigned from civil police, military police, or other troop units specially qualified for security duty. There will be greater economy of resources and efficiency when units are attached to a particular railroad organization or division for the specific purpose of providing security for railroad operations. The guard force on a freight train should be concentrated in one or two positions and should have radio communications with friendly units in the area that could provide support in the event of ambush.

Air cover should be supplied to the train, at least sporadically, and reconnaissance along the right-of-way should be made at irregular intervals.

Aboard the train, automatic weapons should be mounted to deliver fire along the right of way as well as into adjacent areas. When passing through areas of likely ambush—such as ravines, defiles, forests, or areas overgrown with heavy underbrush—token automatic weapons fire may be used to adversely affect the morale of potential ambushers and make the friendly forces feel better. The idea is to scare the hell out of any would-be attackers.

But such reconnaissance by fire must be restricted and permitted only on the order of the convoy commander.

For added protection against derailment caused by sabotage and mines, an engine pushing cars loaded with rocks or earth may be run ahead of important trains. It is probably necessary that the engines of all trains have one or two such cars on the head-end as well. These lead cars are effective against mines in the roadbed (unless they are command detonated). The rock-filled cars won't prevent an ambush, and they won't keep a train or engine from all damage. The cars simply take the brunt of the blast; because they are cheaper than engines and much easier to replace, it is a cost-effective measure. A mine will probably wreck the car and may damage the roadbed. At the very least, it is likely that the blast will close down the line until a wrecking crew is brought to the scene.

It is important to keep in mind that the wrecking crew may be the real target of the ambush. The assumption must always be made that any mine blast is a precursor to an attack against the repair crews. If repair crews can be ambushed and either killed or terrorized into refusing to work, the entire rail system can be brought to a halt just by normal mechanical breakdowns.

Guard posts may be established at critical installations and rail facilities such as tunnels, bridges, and stations. Doing so, however, invites raids against the post itself and ambushes against the troops when they are on patrol away from the post.

Security detachments guarding the right-of-way should have their own communications system, one tied in to the administrative communication system of the railroad.

There are many guerrilla forces in the world that are competent when it comes to railroad attacks, but the Khmer Rouge were past masters at the railroad ambush. On July 1, 1990, at Kampong Trach, they halted a passenger train in an ambush, then executed some of those aboard. At least 26 people were reported killed. Two weeks later, on July 15,

Khmer Rouge effectives slaughtered at least 30 people in a railway ambush at Khampung Chnang.

Indian insurgents also have a good record against trains. For instance in October 1990, Sikh separatists ambushed a troop train near Ferozepur, Punjab. They used a bomb to derail the engine and the six head-end cars, then fired on troops as they tried to disengage themselves from the wreckage. Five troopers died, and 15 were wounded.

Studies of Khmer Rouge attacks or the Indian attacks are worthwhile for anyone charged with defending rail lines against ambushes.

RIVERINE AMBUSH TECHNIQUES

Rivers and sloughs make an effective killing zone. The riverine environment quite often lends itself to attacks on small craft, whether they are insurgent "blockade runners" or patrol boats cruising the waterway.

Ground forces positioned along riverbanks and patrol craft concealed at the edge of a waterway can carry out ambushes. Boats can also be used to put an ambush force in place, withdraw the ambushers from the area, or carry out rapid pursuits. Ground forces that are transported by boats do not necessarily have to conduct waterway ambushes. They can debark from the boats and stage ambushes in any ground area accessible by water. When the ground-based ambush is designed to cover a road, trail, or nearby waterway, the force normally debarks and takes up concealed positions. Boat crewmen remain in or near their craft, which have been carefully concealed. The ambush security leader is made responsible for security of the boats; boat crewmen are under his control during the occupation of the ambush site.

Stealth in movement along waterways is key when troops are being carried by boat to an ambush site, whether that site is on or off the water. Instead of motors, paddles or poles may be used to propel an ambush or insertion craft. The tidal flow and current can also be used to quietly drift the boats and those aboard into the proper position. When paddles, poles, drift, or current are used, however, the craft's motors should be ready so that they can be used immediately if a target appears or the boats are compromised or attacked. Where watercraft stop frequently and debark and then re-embark troops, a stay-behind force can also be left. However to be effective, the stay-behind ambush force must be relatively small in comparison to the force on the boats.

Several hours of waiting are generally required at any riverine ambush site. During this time, there may be changes of the water level and even the direction of flow of the river or stream. Ambush commanders have to anticipate such changes and plan accordingly. For instance, water-level changes caused by tides may require that weapons be repositioned because of alterations in fields of fire. The direction of approach of the enemy craft may be based on the direction of the current. In an ebbing tide, waterway withdrawal routes may become too shallow to use, or craft may actually be stranded—left high if not necessarily dry. Tidal effects can also make landings across mud flats difficult, affect operations around bridges, and leave some weapons useless for periods of time.

When planning, keep in mind that many small streams in the area of operations (AO) may be navigable only at high tide. This fact is particularly important when planning troop landings and identifying support stations.

It is also well to remember that tides and currents are frequently the primary determinants of speed of advance (SOA) of a vessel. Fighting a 3-knot ebb current en route to the AO will significantly affect the speed, and the ebb tide can have a major effect on whether the unit can be infiltrated under cover of darkness. Conversely, utilizing a three-knot current to proper advantage will cut down transit time significantly.

In many areas, high tide is the only time to conduct landings from small boats. If conducted at or near low tide, troops are required to plod through mud and muck that frequently can be waist deep.

At ebb tide, boats may be stranded or some withdrawal routes may become too shallow for use. In one area of Vietnam there was an average 10-foot tidal change. That allowed transit of numerous small streams at high tide, but extreme caution had to be used by the unit commanders to avoid being trapped by a rapidly falling tide and fast currents.

Tidal changes can also have effects on the operations that require boats to pass beneath bridges. For instance, bridges severely limited the accessibility of the Mekong Delta waterways to riverine assault craft. Frequently, however, passage could be effected with caution at low tide.

Care must be taken in the beaching of boats in the AO. With a rapidly receding tide, a boat may find itself aground with no possibility of being refloated until the next high tide.

When troops are being transported by boat, it is important to make certain that people and equipment essential to carrying out the mission are not all loaded into a single craft. If one craft is lost because of a grounding, mechanical breakdown, or enemy fire, the commander of the ambush force must have sufficient means to carry out the mission. Although it is never wise to put all your eggs in one basket, unit integrity should be maintained.

There should a "bump plan" that outlines the minimum force necessary to accomplish the mission. The bump plan also provides the pattern for the redistribution of critical assets if there is an equipment failure or some other problem that prevents the entire force from completing the mission.

When an ambush force is being transported by water,

the insertion point should never be in a position where the craft will have to pass through the area that has been designated as the killing zone. In addition, the boats should be used to provide security at the insertion and extraction points. In most cases the two points should not be the same. A good plan always calls for separate insertion and extraction points, where that is possible.

Gunboats can be used to support the ground force, but detailed fire-support plans need to be drawn up so that friendly-fire casualties do not result. Gunboats and other riverine assault craft can be used by the ambush commander to secure danger areas or to provide flank security. When used in such role, however, the boats should be given flexibility in planning and carrying out their supporting roles.

Occasionally boats, without any assistance from or involvement of ground forces, may carry out ambushes along a river or stream. The general principles found in land-based ambushes still apply to these maritime attacks. However, there are some special considerations that are unique to riverine ambushes.

First, a number of factors must be taken into account when selecting an ambush site for boats:

- Depth of the water at the ambush site
- Obstacles in the water, going to, coming from, and at the ambush site
- Tidal changes
- Weather
- Concealment along the bank
- Fields of fire
- Avenues of approach to the site
- Illumination needs and availability

When a boat orboats will be used to carry out the ambush, the final preparation and checking of gear should be done before moving into the ambush site. This includes weapons, night vision equipment, radios, and so forth. Loose gear and lines need to be secured, snugged down, and silenced. Any spills of gas or oil should be cleaned up, and oily rags should be put into airtight containers. The smell could prematurely reveal the position of the craft, or a slip on them could cause noise and result in a broken ambush.

The craft should be secured with light tensile line or by use of a quick-release device such as a pelican hook. The line should be strong enough to hold the craft against the current, but it should not be so strong that it would hold the craft if the coxswain needed to get under way immediately. Because there may be a need to get the craft under way quickly, it should be staged for a quick get-away, with throttles and helm preset. The bow of the craft should be pointed into the current when it is secured. Appropriate camouflage and concealment is needed.

Fields of fire should be determined early. As each boat pulls into its assigned position at the ambush site, the ambush commander and the commanders of the individual craft should assign sectors of fire and designate rear security. The sectors of fire—both of boats and weapons stations aboard the boats—should overlap. The fields of fire must be strictly controlled.

Rear security should be designated by the ambush commander, and at least one two-man security team should be put ashore from the boats. In the event the team is separated from the rest of the patrol, an escape and evasion (E&E) plan is a must. Contingency plans should include the rescue of the security team and rendezvous points.

ROADSIDE BOMB AMBUSH

This tactic is used against both civilian and military targets, usually vehicles. It is used by both regular armed forces and irregulars. When properly set up, it is extremely difficult to defend against, particularly in cases of executive ambush. In military-type ambushes, claymore mines, with a directed blast pattern, are often used to achieve maximum shock effect.

Most expert terrorists who employ this tactic—either against civilian or military targets—generally use types of explosives that generate great amounts of heat. In that way they get an incendiary effect, as well as achieve blast damage. (See Lebanese Car Bomb Ambush and Armored Cars.)

ROUTE PLANNING

Planning the route from the base of operations to the killing zone is just about as important as silence and camouflage discipline at the ambush site. It can become a matter of life and death. Entrance to the primary ambush site should almost always be made from the rear, and entry into the killing zone should be avoided wherever possible. If the killing zone has to be entered, all telltale signs need to be erased or covered up. When demolitions are being placed or the far side of the killing zone needs to be scouted, detours should be made around the zone. Route planning also involves the selection of an LUP prior to the ambush being set and the designation of rendezvous positions after the attack is sprung.

SIGNALS

Because of security requirements and the noise of battle, the ambushers' command and control system is always subject to breakdown. Any failure or compromise of the signal system will seriously hinder the attacking elements and reduce the effectiveness of ambushes. In order to coordinate the action of the ambush party, the attacking commander must have at his disposal a system of signals to control the actions of the elements of the ambush party. These signals range from the advance warning that the target is approaching to the final signal for withdrawal.

Ambushes need a relatively simple but effective system of signals. Oral commands are generally ineffective because of the noise of battle, or unusable because of security requirements. Hence, makeshift signals, often hand signals, are generally employed. A whistle or air horn can sometimes be used effectively, particularly on smaller ambushes. The use of cords, strings, or vines to connect the members of an ambush party, with a certain number of tugs on the string indicating the message, is another common signal system. It is called, for obvious reasons, a tug line. Pyrotechnics such as flares or colored smoke can be used in some cases.

In a few cases, using advanced technology that allows people to speak over a radio system at a whisper, ambushers can use a radio network. Radios with "bonephone" receivers make no discernible noise, and "whisper mikes" require virtually no sound. But this is a technique that only Special Operations forces have been able to consistently use effectively because of their access to high-tech, state-of-the-art communications gear.

Where possible, an ambush commander should set up redundant signal systems—one to back up the other. The failure of a signal system can wreck an ambush and lead to the death or capture of all the ambushers. When setting up communications systems at an ambush site, keep in mind the Navy SEAL adage that "one is none and two is one." In other words, it is better to be safe than sorry.

SILENCE

Ambushers and units on patrol need to think and practice silence. Silence, in both commands and movement, is important all the time. Given practice and experience, troops can move at high speed relatively quietly.

At the ambush site, silence is an absolute must. The slightest sound can upset an otherwise-perfect ambush. How slight? Studies have shown that the main cause of ambushes being sprung prematurely is the release of safeties on weapons. There are different theories on when to release safeties. Some say they should be clicked off when the ambush party first settles in. Another school of thought holds that they should be taken off at the moment when the ambush commander fires his first rounds. Another view is that they should be clicked off when the ambush target is still some distance away. All have a certain amount of validity, depending on the type of target, the speed at which it is moving, and the weapons' handling ability of those in the ambush party. What is clear is that careless handling of weapons once safeties have been removed can cause disaster.

SIMULTANEOUS AMBUSHES

In favorable terrain and during periods of low visibility, an attacker may simultaneously ambush forces moving toward one another. After deceiving their enemy into a pursuit toward the other ambushed unit, the ambushers pull back and leave the two enemy units fighting each other.

SINGLE-VEHICLE AMBUSHES

Any organized guerrilla force has the ability to ambush single vehicles. Lone gunmen can even do it. Because those in a single vehicle generally lack both adequate firepower and maneuver forces, the best course of action for them is to attempt to drive through the ambush site if at all possible. If this is impossible because of some physical roadblock, such as a downed tree, the vehicle should then attempt to escape using the route over which it entered the ambush site.

Single vehicles are an open invitation to ambush. Similarly, small convoys, particularly ones without visible vehicular communications, almost beg to be hit. The presence of a jeep or two with conspicuously visible antennas is a deterrent to this type of ambush. If nothing else, it implies the possibility of troop support immediately following the ambush or the early availability of air support. When it is necessary to use small convoys (ones with as few as three or four vehicles), one trick is to make the element appear to be a part of a larger body. Suggesting that there is more to follow or that the convoy has a capability that it really doesn't may reduce the likelihood of it being ambushed.

SMELLS

Some people swear they can smell an ambush. In some cases they can. Anything that gives off an odor that would be out of place in the setting of the ambush is a possible cause of a broken attack. Be conscious of any unusual smells such as chewing tobacco, urination and defecation, and colognes or lotions. After-shaves and colognes should be avoided. Besides giving the enemy an olfactory warning that ambushers are near, they often attract bugs and insects.

In Vietnam, the body odor of many Americans, and just as often the deodorant and soaps they used, were detectable by the VC. The lingering smell of cigarette smoke on uniforms and equipment is another possible tip-off. Tobacco products, whether smoked or chewed, give off distinctive smells that betray an ambush position. (In addition, nicotine has an effect on body chemicals and reduces the ability to see well at night.)

Food should never be cooked at an ambush site; the smell of the food is both recognizable and persistent. Food smells that waft around when ambushing forces eat in place are easily detectable by observant scouts.

When using field-expedient camouflage materials, it is best to avoid fuel, oil, and greases to color cloth since they often have a lingering and distinct smell.

Although odor-neutral insect repellent should be

slathered on all members of the ambush party liberally, aerosol sprays should not be used. The cloud of mist has a tendency to travel for long distances and can alert the target.

SNIPING ATTACKS

Snipers serve several roles in ambush situations. In executive ambushes, a sniper may be the entire ambush party, killing the sole target with a single shot.

Snipers can also be used as a mini-ambush in conjunction with mines. They can cover a mined area, waiting for enemy forces to stumble into the minefield, then attack individual combat leaders in a quick harassing ambush.

Sniping is an interdiction technique. Economical in the use of personnel, equipment, and ammunition, sniping has a demoralizing effect on anyone ambushed. For instance, sniper ambushes as used in Northern Ireland by the IRA have had an effect completely out of proportion to the cost of mounting the attack. In 1993, large numbers of men, as well as aircraft and other assets that could have been used elsewhere for road patrols, searches, and traffic blocks, were committed to antisniper operations because of snipers in Ulster "bandit country."

A few trained snipers can cause casualties, deny or hinder the use of certain routes, and force the opponent to use disproportionate numbers of troops or security personnel to keep them at bay. Snipers generally operate best in teams of two, alternating the duties of observer and sniper.

SOLDIERS AFOOT

When terrain and other similar-type cover is available, dismounted personnel caught in an ambush should take cover, set up a base of fire, maneuver their forces, and carry out a counterattack by fire and maneuver tactics.

STAY-BEHIND FORCE AMBUSHES

During search-and-clear operations, search forces sometimes establish ambushes in areas where the enemy is expected to return after the operation is over. These may be ambushes of opportunity or deliberate ambushes. Stay-behind forces can also be used in the riverine environment, where boats transport troops and frequently disembark and re-embark them.

TAILGATES

Tailgates on vehicles should always be left down so that troops can immediately detruck. When in the up position, tailgates slow down anyone trying to get out. When microseconds count, a tailgate in the wrong position can be a death sentence.

TARGET SELECTION

Planning of ambushes or other military/paramilitary operations is essential. The better-planned a mission is, the more likely it is to be a success and the less likely to cause friendly casualties. Target selection is a key part of planning.

There are many ways of selecting targets and determining whether a particular target is the best one to attack. The CARVER system is one way of rating the relative desirability of attacking potential targets and allocating what are always limited resources for subsequent operations. CARVER is an acronym that stands for:

Criticality
Accessibility
Recuperability
Vulnerability
Effect on Populace
Recognizability

Each potential target is rated on a numerical basis, usually either 5 or 10 points, in each of the CARVER categories.

Criticality

Criticality is the initial question. How important is this target to the process? How quickly will the impact of the attack affect other operations and essential systems? How essential is it? Does it have such military, industrial, economic, or political potential that loss of it would severely hamper the enemy? The ambush-killing of a high government official, as an example, would generally have more effect on a government than an attack against a squad of soldiers on routine patrol.

Accessibility

Accessibility is the next issue. Can the target be reached successfully? It may be that the squad of soldiers is reachable, but that high government officials are not. Assessing accessibility means considering the critical path:

- Movement from the staging base to the target area
- Movement from the entry point of the target area to the target itself
- Movement into the target's critical element(s)
- Exfiltration and return to the staging base

Recuperability

Recuperability comes next. That is an estimation of the amount of time it will require to replace a person or critical component or repair damage.

Vulnerability

Vulnerability is a fourth factor that must be considered. This is not the same as accessibility. The government official might be accessible—an attack team could see his car driving back and forth from work daily—but it is an armored vehicle and the attack team doesn't have any weapons that will penetrate armor. This question is most often wrapped up in issues of available assets and technology.

Effect on Populace

This factor relates to whether the act will have a positive or negative influence on the community, the people nearby. An ambush might have either a positive or negative effect on the population often it will be both. A major question is whether the attack will alienate the government from the people, or will it have the opposite effect? In places where there will be reprisals against the civilian population, the effect is a major consideration. Marshal Tito, in Yugoslavia, deliberately staged ambushes knowing that there would be reprisals against the local population, thus forcing the people into his camp.

Recognizability

Recognizability is the sixth and last consideration. Simply put, can the target be recognized? In a prisoner ambush, can the person, or people, to be abducted be recognized so that they can be isolated by fire without being killed and finally captured. Time, weather, terrain, and geography have major effects on this. Nighttime, for instance, often makes recognizability more unlikely.

The points range from one to five, or one to 10.

1 = Bad (opportunity for disaster)
2 = Poor
3 = Fair
4 = Good
5 = Excellent (opportunity knocks loud)

In the example below, five is in your best favor; one is definitely against you. In this example, possible targets being considered for an ambush include:

- Foot patrols of troops
- Armored military patrols
- Province governor
- Village headman
- Suspected informant
- Villagers

When put into the CARVER matrix, the numbers might appear like this:

	C	A	R	V	E	R	Total
Foot patrols of troops	1	3	1	2	2	4	13
Armored military patrols	1	3	1	1	2	4	12
Province governor	5	2	3	1	4	3	18
Village headman	3	4	4	4	5	4	24
Suspected informant	4	3	5	4	4	4	24
Villagers	1	5	1	5	1	3	16

An ambush against a village headman or a suspected informant is the smartest move in the case cited here.

The CARVER matrix is most useful when it is applied to systems and sub-systems and when methods of attack are not limited to one single method, such as ambush. When used in its most strategic way, for instance, it would throw open all systems—governmental, military, economic, political, social, etc.,—to all forms of attack, not just ambushes.

TERRAIN-FEATURE AMBUSH

Experts in the field note this type of ambush was used in one of the earliest recorded instances of bushwhacking—the closure of the Red Sea on Egyptian troops as they pursued the Israelites. Most experts also generally agree that it would take an act of God to make this tactic work. Terrain feature ambushes generally are "ambush by avalanche" or some other diversion of water, liquid or frozen. Creating a rock slide or an avalanche above a convoy or movement of troops below has, in theory, the potential to cause serious damage to the target. For the most part such ambushes pose more danger to the ambushers than the ambushed since these attacks are often mounted from exposed positions. There is little way of controlling the slide or avalanche to make certain it covers or creates a killing zone, nor even to assure that the slide will extend as far as the intended killing zone. A field expedient at best, the terrain-feature ambush looks better in the movies than it works in real life. It is seldom used outside of Hollywood back lots.

TIME FRAMES

Ambushes should take only the minimum time necessary to accomplish the mission. The VC were experts at timing. Their ambushes normally occurred in two phases. A one- or two-minute period of intense small arms fire came first, immediately followed by assault of the ambushed vehicles to complete the killing, destruction, and looting of desired equipment. The always-violent assault phase varied in duration but was usually held to the minimum time required. The fury of the Vietcong attack was as tactically sound as the timing.

TRUCK GUARDS

Military men in troop-carrying vehicles must be

constantly alert and prepared for immediate action. In convoys or individual trucks, selected individuals should be ticked off as special lookouts or guards. In large vehicles, four men should be posted—two at the front and two at the rear of each vehicle. Each is assigned an area of observation covering 90 degrees, from the center of the road to the side in every direction. The truck guards should be armed with automatic weapons, as well as fragmentation and phosphorous smoke grenades. Phosphorous smoke grenades are particularly useful as an anti-ambush weapon. If ambushed, the truck guards fire long bursts to cover the detrucking of the other troops. The guards may also aid in control of the convoy, short of an ambush, by informing the vehicle commander when the vehicle following drops back or stops.

ULSTER CAR BOMB

This is a terrorist tactic in which an explosive charge is planted on, inside, or underneath the vehicle. It is used largely in ambushes aimed at executives and other protected persons. Despite the name, the car bomb is used throughout the world. It is not a phenomenon limited to Northern Ireland. In the United States, Mafia car bombings are often of this type.

The best, in fact about the only, way to defeat this type of ambush is to either prevent the bomb from being emplaced or to detect it before it goes off. The first line of defense is a good car alarm system. The second defense is a search of the vehicle for bombs—one before every use. The purpose of searching a vehicle for bombs is simple—to get enough information to decide whether there is a reason to call the security and explosives ordnance disposal specialists.

The searcher is looking for telltale signs that the car has been tampered with in some way. Finding out how it may have been tampered with, or whether there is any danger because of the tampering, is up to experts.

The actual process of bomb checking starts the day the vehicle is acquired. Color pictures of the engine compartment, shot from at least three different angles, should be taken. Pictures of the underside of the car are also useful.

To make the bomb-check process itself safe, the vehicle should be garaged in an access-controlled area. An area that isn't access-controlled makes it far too easy for someone to plant a bomb. Even more importantly, checking a vehicle for a bomb in the open makes a person extremely vulnerable to other types of ambush. Checking a vehicle for a bomb takes a few minutes—minutes of intense concentration. No one is ever more vulnerable to an assassination, a kidnap attempt, or simple gratuitous violence than he or she is while walking around the car as it sits parked on a dark street.

Generally, it is easiest to start the bomb search by making three circuits of the vehicle.

The first walk-around covers, not the car, but the ground surrounding the car. The idea is to look for pieces of tape, bits of (often shiny) cut wire, shunts from electrical detonators, and lengths of fishing line—the last used to set off a booby trap on the walk-around. Since car inspections quickly become a matter of routine—and will undoubtedly be done in virtually the same way every time—it's important to watch for trip wires and other booby traps set up to catch the jaded, weary searcher. During this phase of the inspection, you're also looking for scuff marks on the ground or concrete. These will often show up where someone has crawled under the car or his or her shoes have dug into the ground.

The next circuit involves a visual inspection of the upper surfaces of the car—the roof, trunk, and hood. Evidence that suggests tampering includes paint scratches, smudge marks, or clean areas on surfaces that are soiled by road dirt. These "clean marks" could indicate that the doors, hood, trunk, windows, or sun roof have been forced open. For those people who are really into the checking and carry a small container of talcum powder to sprinkle lightly on the door handle when they leave the car, this is the time to check for the residue. A small piece of cellophane tape on the seam between the hood and fender, or between the door and frame, will break if someone has gotten into the car. This is also the time to check for those breaks if you use either of those techniques.

The third pass around the vehicle is the time to inspect the area between the door panel and the ground. Special attention should be paid to the hubcaps to make certain they have not been taken off and replaced. If a hubcap has been taken off, there is no reason to try to find out what, if anything, is there. It could be something as innocuous as a rock placed there by a precocious neighbor child. It may be that all the lug nuts have been taken off but one. The hubcap might be harboring an explosive device—one designed to catch the curious. The wheel wells should also be checked thoroughly. During the third pass, it's important to look under the car, from the engine compartment to the gas tank. The inspection should be done from the front, sides, and rear. The idea is to check for any canisters, wires, or other protrusions—especially around the exhaust and gas tank. Grease smudges and dirt, anything that wasn't there before, are indicative of a potential problem.

The gas tank area should be checked carefully. The car should have a locking gas cap, even if the gas cap door can (supposedly) only be opened from inside the car. (Metallic sodium, placed in a gelatin capsule and dropped in the tank, will eventually blow the car to flaming bits.)

The car should have a locking hood. That means that in order to continue the inspection it will be necessary to get inside the car. Reinspect, carefully, the driver's side door and the car's interior to see if anything is out of place. That process can be aided by leaving a box of tissues, a book, or a similar object on the front seat. This marker should be left in the same relative position and orientation to the seat each time the car is parked. If it has been moved or if the orientation of the marker has changed, there's reason for you to be concerned.

When satisfied that no one has booby-trapped the car door or installed some sort of motion-sensor trigger, the car door can be opened and the hood unlocked. But before actually opening the engine compartment, the hood area and the hood-opening mechanism should be inspected again for obvious signs of tampering.

When the hood is opened, you look inside for bombs. Some mechanically minded people know what everything in the engine compartment does. For them, it's an easy inspection. They'll notice any additional wires or extra "components" immediately.

Most people aren't that attuned to what happens under the hood. That's where those pictures taken earlier come in. Compare the pictures with the present view. Special attention should be paid to the area around the fire wall, between the engine and dash. The fire wall, immediately in front of the driver, is a favorite place to put bombs. Things that should trigger suspicion during the under-the-hood inspection include greasy parts on a spotless engine, shiny new parts on a dirty engine, or a part that wasn't there before. If those things are apparent, there's no need to even close the hood. The best move is to leave and call for assistance.

If the engine compartment checks out as safe—and it almost inevitably will—a quick check under the front seats and the dashboards for other unwanted "accessories" is wise. The Red Army Faction, for instance made use of pressure devices under seats; the technology is well known in the terrorist world. It's also wise to look carefully at the rearview mirror and the headrest. Some groups are trained to place explosives such as Flex-X or C-4 under the headrest or inside the rearview mirror. The mirror bomb works off a hearing-aid battery and a mercury switch. Just twist the day-night lever and the C-4 and ball bearings end up in your face!

When everything checks out, it's time to drive. Short of tearing the car apart in order to check the guts of every component, there is little more to be done.

But to where will the car be driven? For protected persons or people likely to be targets of an executive ambush, the destination should have secure parking. In most cases if the car can't be left in at least some sort of semi-security and out of access from the general public, it's better to use some other way of getting around. That's what friends and cabs are for.

The variety of ways to rig a bomb on a car is almost endless; the process of checking for bombs seems almost as endless. But remember the simplest ambush ruse is as deadly as the most intricate bomb. In Tehran, terrorists punctured the gas tank on a car, soaked the surrounding area with gasoline and put a burning newspaper on top of the car. Then they woke up the American owner to tell him that the car was on fire. He ran outside of the house in pajamas, knocked the burning paper to the ground, and set himself and the vehicle aflame.

WARSHIPS AND WARPLANES

There are some, particularly blue-sky jet jockeys and blue-water sailors, who will argue that ambushes cannot be set up in the air or at sea. They insist that ships and planes fight as they move, that aircradft and ships do not stand still and wait for an enemy to come to them. Admittedly, there is some roon for debate if you're a purist. They also contend that the cover and concealment needed to carry out an ambush simply do not exist 600 miles from the nearest land or 40,000 feet in the air. This argument is provably wrong.

"Coming out of the sun" or hiding in the clouds to

pounce down in a surprise attack on an unwary enemy is a well-known aviator's trick that has been in use since planes first rose into the wild blue yonder. On the ocean, the submarine and the armed merchant cuiser use the concealment that their particular craft offers. Submarines enjoy the ability to wait in ambush below the level of the sea. Armed raiders employ the element of surprise afforded by a false flag and heavy guns mounted on what appears to be a peaceful merchant ship.

In more recent times, it has been common for warships to use islands to mask radar, then steam out to attack suddenly from the cover of the island hills. While it is hard to argue that aviators or sailors speak in terms of ambushing their opponents—they simply don't talk in those terms—they do carry out what amounts to ambushes. Sometimes these are on a massive scale.

Adm. Isoroku Yamamoto, the architect of the Japanese air raid against Pearl Harbor and a top-flight naval tactician in many ways, seems to have had particularly severe problems countering the ambush. Because American cryptographers had effectively unraveled major parts of the Japanese code, Yamamoto's thrust to take Midway—a plan in which Japan expected to lure and destroy in detail the smaller U.S. fleet—was both known and clearly understood. A feint at the Aleutians designed to get Adm. Chester Nimitz to split his fleet was seen for what it was, and the U.S. Navy laid a watery trap for the Japanese.

Yamamoto and his fleet were traveling blindly toward Midway, like an army column without scouts or flankers along a well-mapped road in friendly territory. But his general route and destination were known to the U.S. Navy, which had moved up to defend the mid-Pacific bastion. As Yamamoto's carrier forces attacked Midway on June 3, 1942, the Japanese admiral was shocked to learn that there were unidentified U.S. ships (they turned out to be the carriers *Hornet, Yorktown*, and *Enterprise*) not far to the northeast. The Americans were in a position where they had no right to be, based on all of Japan's intelligence.

Yamamoto's carriers were in the killing zone of some extremely long-range weapons—U.S. Navy carrier planes. The Japanese carriers *Akagi, Soryu, Kagi*, and *Hiryu* went down in the battle. But the United States lost as well. Much of the U.S. aerial fleet was destroyed, and the carrier *Yorktown*, damaged by bombers, was finally sunk by a Japanese submarine, I-168, lying in watery ambush. But the ambitious U.S. ambush of the main Japanese fleet dealt Nippon the first—though far from the last—naval defeat in its history.

Yamamoto himself died in an aerial ambush less than a year later. U.S. intelligence had deciphered messages in the JN25D code giving details about a morale-building tour that Yamamoto was taking of front-line bases. Sixteen U.S. Army P-38 fighter pilots—briefed on the hour, location, and even the escort of his inspection flight—timed their arrival at Buin in the Solomons. They chose the moment when Yamamoto's "Betty" was about to land, and then they pounced. The commander who was planning to meet his troops met his death instead. The P-38s caught the two VIP-carrying Bettys and three of the nine Zero escorts in a withering fire that sent them crashing to earth. But even successful ambushes cost the attackers. A U.S. Army Lightning was lost to return fire.

On the other side of the world, the German armed merchant cruiser *Atlantis* had come to grief earlier in the war, but not before she had sunk or captured 22 ships with her hidden guns. Posing as a neutral ship or a vessel of Allied nations, *Atlantis* approached hapless merchant ships that came marching across the horizon, then flayed them with gunfire if they did not surrender. The British, in a succesful effort to catch her and her cohorts, staked out the areas where calm weather conditions made it most likely she and U-boats would be refueling. And eventually, like a truck convoy rumbling down a well-traveled road, the *Atlantis* and members of her oceanic entourage rolled into the trapon November 22, 1941. She had been ambushed like the ships she had sunk.

WATERWAY TRAFFIC AMBUSHES

Waterway traffic such as barges, ships, gunboats, and other craft may be ambushed just as simply as a car or column of military vehicles. The ambush party may be able to mine the waterway and thus stop traffic. If mining is not feasible, recoilless weapons fire can damage or sink the craft. Such fire is most effective when it is directed at engine room spaces, the waterline, and the bridge. Strikes on the superstructure, other than ones on the bridge and radio shack, are virtually worthless.

Usually waterway ambushes are either designed as killing or harassing ambushes, although sometimes the attack is designed as a resupply ambush. Recovery of supplies may be possible if the craft is beached or grounded in shallow water.

Personnel who operate waterway facilities, such as pilots and lock operators, are key targets for land-based killing ambushes. These personnel are not easily replaced. Like track repair crews, they are essential to keeping the transportation system working.

WEAPONS OF AMBUSHERS

People can use a rock or a pen knife to carry out an ambush. Palestinians, for instance, have used these when attacking civilians in the Occupied Territories. But these are never the weapons of choice, for obvious reasons.

Weapons must be carefully chosen for an ambush. Each type of weapon has its own strengths and weaknesses that makes it useful, or rules it out, for the particular attack.

The VC, who mastered the ambush and even now are pointed to as the experts, used small arms as their basic ambush weapon, augmented by automatic weapons, and on occasion by recoilless rifles or rocket launchers. Mines, which were almost always electrically detonated, were employed to cause personnel casualties as well as to disable vehicles. At the outset the mines consisted of altered 105mm artillery shells, mortar rounds. Later conventional mines were brought on line.

Modern terrorists carrying out ambushes against civilians and officials tend to use small arms and automatic weapons, as well as roadside bombs.

Aircraft ambushes can be carried out with Stingers or large-caliber sniper-type weapons. Ambushes along inland waterways often involve automatic weapons, mines, and recoilless weapons.

WEAPONS PLACEMENT AND USE IN CONVOYS

In convoys, automatic weapons should be placed so they can be fired at ambushers immediately. Standard or improvised mounts, constructed so that the guns can be quickly removed from a truck or vehicle, are recommended. Crew-served weapons should be distributed throughout the length of the convoy to provide indirect fire support. In an attack, these can easily be removed from a vehicle and rapidly placed into a firing position, or they can be fired from a vehicle bed. Troops with rifle grenades and grenade launchers should fire those weapons immediately on contact. Phosphorous grenades are particularly effective, for they not only produce an immediate and effective screen but are also a feared casualty-producer.

WEATHER

The elements can affect an ambush. Both sides try to take advantage of weather factors. For instance, rain covers the sound of movement. The dampness of the ground or the vegetation muffles noise for those working in the jungle.

Wind direction and velocity can be important considerations in choosing an ambush site. Whenever possible, the killing zone should be located downwind of the probable direction of the target's approach. The sounds carried by the wind give the ambushing force advance warning of the enemy's movement toward the killing zone. The wind also makes it harder for the enemy to detect the presence of an ambush ahead because it masks sounds and carries off odors from the ambush site.

WIDE-AREA WEAPONS

Experience has indicated that the best way to minimize casualties when caught in an ambush is to escape from the killing zone as quickly as possible. Wide-area weapons can be useful in doing so.

To escape the killing zone as quickly as possible, the books all say, "The reaction of the ambushed force must be violent in nature, characterized by automatic action and employment of weapons that have a large area lethal effect." That means use every big weapon in the arsenal and throw everything at him that will hurt. Examples of the wide-area weapons are grenades (hand, rifle, and M79), flamethrowers, and 57mm recoilless rifle canister rounds. Units on the move should always have such weapons ready to deliver area fire or fragmentation effect.

3
Case Studies

There are lessons to be learned from others' experiences. The following case studies illustrate how ambushes have been used around the world in various situations. The reasons for their success or failure are also noted.

QUANG TRI AMBUSH (VIETNAM, 1960s)

Many lessons are apparent in this study of a large-scale Vietcong ambush of a Vietnamese Marine battalion while it was moving from Hue to Dong Ha by motorized convoy. A classic example of an ambush, it cost the Vietnamese Marines 137 casualties. Prompt reaction to the ambush by the battalion survivors, as well as the speedy arrival of a reaction force, prevented a serious defeat from becoming catastrophic.

Available intelligence showed that no major contact or ambush had occurred along the line of march in the previous 10 months. Other than the usual concerns, the troops—who were going to Dong Ha by way of Quang Tri City—had no particular jitters.

The weather was hot; temperatures ranged from 85 to 90°F during the day. The skies were clear and visibility was excellent. The terrain was generally open. Rolling hills were interspersed with short lines of shrubs and thin stands of trees. Vegetation was heavier northeast of the road, with a dense tree line 1,000 feet away.

At the point where the ambush took place, the observation and fields of fire could hardly have been better. Southwest of the road, observation ranged up to 1,500 feet with excellent fields of fire. To the southeast, observation from the road extended approximately 250 feet to the top of a slight rise, which masked the terrain farther east. Excellent fields of fire were available for all weapons emplaced on that rise.

The Vietnamese Marines were not without support. A battery of 105mm howitzers and one battery of 155mm howitzers of the 12th ARVN Field Artillery Battalion were to support the move in its initial stages. Initially, the plan was to have a U.S. L-19 aircraft on station for the visual reconnaissance and communication assistance, but just prior to the jump-off this support was canceled and replaced with a Vietnamese Air Force L-19, which joined the convoy about three miles north of Hue.

Troops were loaded onto the vehicles by company. The order of march was 1st Company, 3rd Company, H&S Company, and the Command Post (CP) Group, 2nd and 4th Companies.

Personnel were required to face outboard and have weapons at the ready. Artillery fires were planned along the route of march, and an AFO team was attached to the battalion.

The battalion commander's antiambush instructions to battalion were to dismount, form up by units, and stand to fight as he directed.

The battalion crossed the LP at 7:30 A.M. and moved toward Quang Tri City. At about 8:30 it entered the ambush site.

The L-19 aircraft had failed to detect the ambush, and the VC, in battalion strength, opened fire. The initial blast of withering fire included heavy small arms and a heavy volume mortar and recoilless rifle fire. The convoy came to halt immediately as three trucks were struck. The convoy deployed along the side of the road. There was little cover and concealment there. As the VC fire improved in accuracy, casualties started to mount.

From the position of the beleaguered battalion, the back blast from the recoilless weapons was easily seen along the crest of the low, rolling hills in the southwest. Small groups of VC could be seen as they maneuvered toward the road.

At that point the battalion commander ordered the rifle companies and the Command Post group to retire to the relative sanctuary of a railroad cut, about 75 yards northeast of their location on the road. No fire had yet been received from that area. The move to the railroad cut was carried out simultaneously by all companies. But as the units closed on the position, H&S company, Command Post Group, 2nd Company, and 4th Company were hosed with a withering volume of small-arms fire and hand grenades. There were an estimated two companies of the VC in well-camouflaged positions along the southeastern side of the railroad cut. In the initial fusillade, the battalion commander was mortally wounded, and virtually the entire Command Post group was killed outright or incapacitated by wounds.

In the vicinity of the decimated Command Post group, about a score of marines had managed to gain the railroad cut. They were establishing a perimeter when they were struck by machine gun fire directed down the railroad tracks into their left flank. That was followed by a vicious infantry attack from a company-size VC unit from the northwest. All the marines holding the cut were either killed or wounded.

The successful completion of this phase of the enemy plan left the Command Post decimated—and inoperable.

Approximately 75 yards southeast down the railroad cut, the remainder of H&S company had formed into a defensive perimeter. This group was engaged in close combat with an estimated VC company that was positioned 20–25 yards to the northeast. After the successful VC assault on the command element, this position became untenable and H&S Company pivoted south and fought its way to the 2nd and 4th Companies. During the fighting dash, H&S Company took casualties from mines laid along the railroad cut.

While all this was going on, 1st and 3rd Companies had formed a joint perimeter elsewhere and were establishing a heavy base of fire in a 360-degree arc.

At this point—about five to six minutes had elapsed since the first contact—the VC had succeeded in dividing the battalion into two separate forces about 500 yards apart. The defense perimeters were not mutually supporting, and, because of the heavy casualties already suffered, the surviving commanders decided to stand fast and wait for reinforcements.

About five minutes after the ambush was sprung, a U.S. L-19 arrived overhead with a Vietnamese artillery observer aboard. Artillery fire against the enemy concentrations began about 10 minutes later. Shortly thereafter, a U.S. Air Force forward air controller arrived and commenced calling in airstrikes. During the entire engagement, aircraft faced heavy antiaircraft fire from weapons placed along the high ground to the southwest.

About 20 to 25 minutes after the start of the ambush, the VC began breaking contact. Enemy units northeast of the road began withdrawing toward the river, and those to the southwest were observed moving rapidly to the west.

Devastating air attacks, coupled with intensive small-arms fire from the two positions of the marines, pummeled the VC as they withdrew, causing considerable casualties.

The excellent air support alleviated some of the pressure on the battalion and steps were taken to link up the separated forces.

About 45 minutes after the Vietnamese Marines walked into the killing zone, a U.S. Marine rifle company arrived and began an immediate and aggressive pursuit of the ambushers. The company moved as far as the high ground on which the enemy had emplaced the majority of its force during the ambush. The company halted but continued its pursuit by fire.

By noon a three-battalion ARVN force, assisted by two U.S. Marine companies, were in the area and had succeeded in trapping the VC. The reaction part of the engagement lasted three days and resulted in 223 enemy KIAs, in addition to the 52 killed by the marine battalion that had been caught in the ambush.

There are a host of lessons to be learned from this ambush. While the VC had, for many years, demonstrated an ability to execute ambushes without premature disclosure of their position, this one was a masterpiece. It was notable because it was conducted in broad daylight by a large force, encumbered with heavy crew-served weapons, and within two miles of bivouacked major friendly units. Effective patrolling and well-placed listening posts would have provided some indication that a large enemy force was in the area.

The enemy commander has to be given credit for carrying out a classic ambush. But he made a pair of glaring errors that resulted in disaster for his forces during the withdrawal phase. First, he miscalculated the amount of time it would take to bring up a reaction force. He waited around too long after the shock and surprise had worn off, trying to inflict further casualties. Second, he failed to provide an adequately covered and concealed withdrawal route for his southwestern elements.

Units withdrawing to the southwest had to move through open terrain for two miles! They presented air and artillery with superb targets. On the other hand, VC units withdrawing to the northeast were afforded a heavy forest canopy and escaped unharmed.

The early decimation of the Vietnamese Command Post group and the subsequent death of the battalion commander left the battalion in crucial straits for about 15 minutes. During that time, the battle evolved into two separate and distinct actions, with loose overall control.

The southern element under the command of the battalion executive officer fought a rugged face-to-face encounter with a dug-in and determined enemy and yet managed to hold its own. The northern element, while out of the primary killing zone and not having to contend with enemy units close-in to their positions, took the brunt of the recoilless rifle and mortar fire in good order. They were able to return a high volume of effective fire.

The key factors in avoiding the disaster that had loomed so large in the first 15 minutes of the engagement were:

1. The maintenance of unit integrity of the rifle companies.
2. The skillful manouvering of these companies by their commanders
3. The downright doggedness, determination, and raw courage of individual marines.

The U.S. Marine advisor was moving with the Command Post group when it came under fire. In the ensuing action he was able to observe enemy troops at close range. They were expertly camouflaged, and all wore the standard fiber VC helmet covered with freshly gathered vegetation. The VC that he could see wore a cape-type garment made of camouflage material. At the arrival of the U.S. FAC aircraft, a platoon of VC were seen to "hit the deck" and arrange the camouflage cape over their backs; they remained motionless. Aircraft flew over them as low as 500 feet and never saw them below!

GAMA'A CAMPAIGN AGAINST TOURISTS (EGYPT, 1992-1993)

Ambushes in war are to be expected, even when they are unexpected. Soldiers are always targets. The Quang Tri ambush is typical in that sense. But foreign tourists don't expect to die in a battle that revolves around niggling points of religious doctrine. Nor does a whole nation expect to be held captive by a handful of terrorists out to destroy the government by wrecking the economy. But that's what happened in Egypt where an ambush-based fundamentalist terror campaign wracked the nation economically, politically, socially, and religiously.

Starting in mid-1992 Muslim militants mounted a highly effective ambush campaign that targeted tourists.

Tourism is Egypt's major producer of foreign exchange and its leading industry. The militant campaign was as simple as it was effective. The weapons—guns and bombs—were simple, effective, and easy to obtain. This inventive ambush campaign, carried out by a handful of people, wrecked the country's tourist industry within a matter of months

The campaign opened in earnest on July 14, 1992, when four tourists were slightly hurt as a firebomb was hurled at a tour bus near Luxor, site of some of Egypt's most famous Pharaonic temples and tombs. On September 30, a spokesman for the main militant group, el-Gama'a el-Islamiya (Islamic Group), warned tourists not to go to Luxor or other sites in Qena Province. The Gama'a warning was followed by more ambush action just a few days later. On October 2, Gama'a attacked a Nile cruiser ferrying 140 Germans on a placid cruise along the river. Three Egyptian crew members were wounded in that waterway ambush.

The boat attack was followed by an even more serious ambush on October 21, 1992. The Muslim militants fighting the Egyptian government ambushed a safari bus in the southern Nile Valley, killing a British woman and wounding two British men. Sharon Pauline Hill, 28, from Camberley, was shot dead in the attack near the fundamentalist stronghold of Dairut, south of Cairo. A boy standing in the road whistled when he saw the bus, and gunmen opened fire from fields on either side. Gama'a, claimed responsibility for the attack—the second on foreigners in 21 days but the first in which any of the travelers had been hurt.

On October 22, Egyptian Tourism Minister Fouad Sultan said Muslim militants who ambushed a bus and killed a British tourist were aiming to challenge the Egyptian government by damaging the $3 billion a year tourism industry. But Sultan played down the attack of the day before, saying it would not affect the country's major foreign exchange earner. "What happened is a regrettable incident," he said. "I'm extremely sad and sympathetic with the victims, but at the same time I want to say that such incidents happen everywhere . . . Fundamentalism has become a worldwide phenomenon. The whole world is highly mature and understands that currents and undercurrents happen. . . . If you have a car accident you cannot stop riding cars. Bomb explosions are hitting London, but that would not mean that London is an insecure and instable country," Sultan said.

Gama'a activists continued to use ambushes in their violent attempt to turn Egypt into a stricter Islamic state. On November 1, 1992, a bus carrying 55 Coptic Christians was ambushed by three gunmen near Deir Mawas, approximately 170 miles south of Cairo. Ten passengers were wounded when the gunmen opened fire from fields on either side of the road. The incident took place close to the spot where the British tourist was killed in a similar incident in October—one in which the gunmen opened fire from a field by the roadside.

Later in November there was another bus ambush when Gama'a gunmen attacked a tour bus in the town of Qena. In the November 12 attack, five Germans and two Egyptians were wounded.

Shortly afterwards, on January 5, 1993, shots were fired at bus carrying 20 Japanese tourists near Dairut. Two days later, on January 7, a Gama'a militant threw a bomb at a tourist bus in Cairo. That marked the first demolitions ambush attack on a tourist target in the capital. The explosion shattered the back window of the bus but caused no injuries to those in the contingent of German tourists.

On January 11, Gama'a again warned foreign tourists to stay out of Cairo and parts of Upper Egypt. Then on February 4, 1993, three Muslim militants threw a bomb at a bus carrying 15 South Korean tourists near the pyramids. The bomb thrown by the militants shattered the window of the South Koreans' bus near the Europa Hotel. The bus was going from the Mina House Hotel near the pyramids to Old Cairo. None of the 15 tourists on board was hurt. Police captured two militants and said they were carrying more bombs in a plastic bag. Police said one of the attackers tried to escape by throwing a bomb at a police car. That bomb exploded in the road, causing no damage. The attack was the second in two months near the pyramids of Giza, Egypt's main tourist attraction. An anonymous caller to an international news agency indicated that the attack had been mounted by the militant el-Gama'a. "A bomb was thrown on a tourist bus on the pyramids road. Another bomb was thrown on the police car protecting the bus," the caller said before hanging up. The man did not identify himself, but followed a normal pattern in that telephone callers speaking in the name of Islamic Group routinely called to claim responsibility for such attacks.

On February 9, militants shot at a bus carrying German tourists near Dairut, but there were no casualties—other than the rapidly failing tourist industry. On February 17, police escorting a group of German tourists shattered an ambush. They shot one militant dead after he and another gunman opened fire on buses near Dairut. Then on February 26, a bomb exploded in a downtown Cairo cafe, killing three people and wounding a score more. The device was apparently intended to be used in an explosive ambush directed against tourists, but reportedly went off prematurely. The explosion in the crowded coffee shop in central Cairo killed a Turk, a Swede, and an Egyptian and wounded 20 people. Gama'a denied responsibility for the blast.

There was no surcease from the ambush attacks. On March 16 a bomb planted by militants damaged five tour buses outside the Egyptian Museum in central Cairo. There were no casualties. Two weeks later, on March 30, a bomb in the pyramid of Chephren at Giza injured two workers.

On April 9, ambushers were back blasting at boats in a revival of riverine ambush. The Gama'a militants fired at a Nile cruiser carrying 41 Germans near Assiut. Nobody was injured. On April 11 a tour guide spotted a man planting a bomb on a bus loaded with German tourists outside Cairo's medieval Citadel and thwarted the demolitions ambush.

On May 21, 1993, fundamentalists successfully staged a Lebanese car bomb ambush, fatally wounding seven people and injuring a score of people on a busy street outside a police station in central Cairo. Three of the fatalities were children in a single family; they had been on their way to a zoo when the afternoon explosion shattered the calm of the city. The 4.4-pound bomb, packed with nails and metal fragments, had been placed in a car parked illegally on a main road behind the police station. It tore a hole in the road and blew out windows in the nearby Civil Registry office, where people record births, marriages, and deaths. The car had been parked in a no-parking zone for 48 hours. It had been parked by a university professor who left it there while he took a bus to his home elsewhere in Egypt. The attackers apparently were able to open the locked car and leave the bomb on the passenger side.

The demolition ambush blew up in what was often a crowded street, one that was not far from the headquarters of two of Egypt's biggest newspapers and Cairo's main railway station. But the street was quieter than usual because it was the Muslim weekend. The car bomb sprayed glass and nail-size shards over a busy downtown district. It was the fifth terror attack in the heart of the Egyptian capital since December 1992. But it was the first time a car bomb had been used in more than a year of violence between Muslim militants and security forces.

On June 8, ambushers set off a bomb on Pyramids Road in Cairo, killing two Egyptians and wounding more than a score of people, including five Britons on a tour bus. Eleven days later, on June 19, Egyptian security forces dismantled a time bomb containing more than 10 pounds of explosives and nails outside a bazaar shop in the southern tourist resort of Aswan, 430 miles south of Cairo. Late in the day, experts defused the explosive device left in front of a tourist souvenir shop on one of the busiest commercial streets of Aswan. Security sources said the bomb, contained in a jerry can connected to a detonator, was made up of a large quantity of the TNT chemical. It was defused only four minutes before its timed explosion. The terrorist bombing ambush was thwarted by a shopkeeper who suspected the plastic container might be something else and immediately reported it to the police.

On July 18, 1993, Muslim militant gunmen, avenging comrades hanged in the previous five weeks on the orders of military courts, ambushed an Egyptian army general's car in Cairo. The Muslim extremists, apparently seeking revenge for the hangings of the radicals, opened fire on a car carrying an army general, then engaged in a series of shoot-outs with police while trying to flee. Police and soldiers exchanged fire with the gunmen at two places, several blocks apart, along a major highway. The general escaped unharmed, but four people died and six were wounded, mostly in the gun battles that followed the ambush near the City of the Dead, a maze of centuries-old tombs and monuments often visited by tourists. One gunman escaped, one was arrested, and two died in shoot-outs, police said.

The official Middle East News Agency quoted Information Minister Safwat el-Sherif, who claimed that Maj. Gen. Osman Shaheen, commander of Cairo's central military area, was not the target of the ambush and just happened to be passing when gunmen attacked a police car. Despite the official denial, the attack on Shaheen was believed to be in revenge for the previous day's execution of five members of al-Gama'a al-Islamiya.

What was clear was that four gunmen fired from a car on a highway, and the shooting took place only about 300 yards from the morgue where the bodies of the five executed men were being returned to their families. One passer-by was killed in the attack. According to the police, the gunmen tried to escape by running into the surrounding slum neighborhood of Sayeda Zeinab, but two of them were surrounded by residents and beaten until police arrested them. One of the gunmen died of his injuries. One other gunman commandeered a taxi but was confronted by police several blocks away. He was shot and killed in a gun battle. The fourth gunman escaped. A policeman wounded in a shoot-out died later in a hospital. Capt. Ahmed Beltagi died in the second shoot-out involving the hijacked a taxi. Apparently the kidnapped driver spotted Beltagi's traffic police patrol car and shouted for help. The fleeing attacker jumped out of the cab and opened fire on the police car and the taxi, hitting Beltagi and another policeman, but was then shot and killed. Four civilians, a policeman, and an army officer were also wounded, police said.

On September 18, suspected Muslim militants ambushed and shot and killed a senior police officer in Aswan. State Security Police Brigadier Mamdouh Osman reportedly was killed by two gunmen who opened fire on his car as he got into the vehicle. The attackers then fled the scene. The same day gunmen opened fire on a Nile cruiser as it navigated through the troubled southern province of Assiut, Egypt's chief stronghold of Muslim militants. Muslim militants opened fire on the boat, carrying 22 French tourists, from positions along the banks but missed it, security sources said. Other reports said the ambushers, who fired from a banks of a plantation, broke one of the craft's windows. The one thing everyone seemed to agree on was that this time no casualties were reported among tourists aboard the ship.

On December 27 of the same year, eight Austrians and a like number of Egyptians were wounded when Muslim militant ambushers threw two bombs and fired shots at a tourist bus near one of Cairo's ancient mosques. Two Austrians were seriously hurt, including a 25-year-old man who was hit in the head. The ambushers threw two explosive devices, one of which exploded inside the bus and the other outside. Despite the blast's effects, the driver drove the bus out of the killing zone, stopping about 500 yards away, and the ambushers fled.

Witnesses said three young men had been waiting for the bus, lounging at a roadside cafe near the Amr ibn el-As mosque. A boy stood up, holding something round in his hand, and hurled it at the tourist bus, causing a commotion. As people in the cafe stood up to see what was happening, one of the trio, with a revolver in his hand, ordered everybody, "Don't move!" He began blasting away at the bus. After the attack the trio ran, pursued by the onlookers, who were then fired on by the men. The Egyptians who were reported wounded were hit either by flying glass from the bus or by shots fired by the attackers. Police immediately closed all roads into the area, a warren crisscrossed by a maze of narrow alleys bordering overcrowded tenements.

A key component of the Gama'a campaign was the repetitive attacks against tourist buses, tourist sites, and Nile cruisers. The ambushes proved just how easy it is to ambush river craft, especially unarmed ones, or tourist buses. Visitors and tourists are looking at the sights; they aren't looking for trouble. Unlike soldiers in a military convoy moving through bandit country, they are unarmed, unaware of the danger, and unable to respond effectively when attacked. Tourists can't fight back; they can only stay away from the country in droves and bring on economic starvation in their fear.

A boatload of tourists can't see or sense imminent danger. Unlike the members of a military unit, they don't have any sense of the danger they are walking into. They are sacrificial lambs. But they aren't the only sacrificial lambs. Sometimes knowing that there is danger isn't enough; sometimes having the training isn't enough either. When political considerations deny a military force essential equipment and when bad planning and poor decision-making compound the problem, there is nothing anyone can do except become a casualty in the killing zone.

BLACK DAY FOR BLACKHAWKS (MOGADISHU, SOMALIA, OCTOBER 3, 1993)

On October 3, 1993, an operation in the Somali capital that was expected to take no more than a few minutes turned into a 16-hour ordeal—one that left about 18 Americans dead and nearly 80 wounded out of a total of about 500 U.S. soldiers involved in the operation.

At about half past 1:00 P.M., the word went out to prepare for an operation aimed at putting pressure on Farah Aideed, the Somali clan chief wanted by the United Nations in the June ambush deaths of 24 Pakistani UN soldiers. The U.S. Rangers, sent to the Somali capital specifically to run the wily warlord to ground, were told to prep for a mission at Mogadishu's Olympic Hotel, a downtown watering hole where top Aideed lieutenants were reported to have gathered.

Approximately 100 Rangers were ordered to fast-rope into the hotel as part of a helicopter assault. They were to grab as many prisoners as possible. Even if Aideed was not among them, his command and control staff would be decimated. The plan called for Rangers in a pair of Blackhawks choppers to secure the hotel and the building across the street, arrest the Somali leaders, and wait for a Ranger ground force driving Hummvees and trucks to roll in. The vehicles were to transport everyone—troops and their prisoners—back to the UN command center at the airport.

Unanticipated gunfire at the landing—which took place about 3 P.M.—suggested Aideed's men had been tipped off. The raiders were being ambushed. But despite the fire, the plan seemed to go as written. The Blackhawks swooped in and let off their passengers; the troopers aboard shut down the Somali shooting by putting up a withering curtain of return fire. The helicopters then sped away as the troop's Special Operations Force took the buildings and about a score of Somali prisoners.

As the Rangers' evacuation convoy approached, the word came in that one of the Blackhawks had been shot out of the sky not far from the hotel—an event that changed the plans and the course of the fighting. The Rangers, under a revised plan, were to reach and cordon off the wreckage and rescue any crewmen still alive.

From that point on, the mission began to go seriously awry. More Somali gunmen surfaced; they set up roadblocks and staged ambushes of opportunity wherever they could. The Rangers fighting their way to the wreckage of the downed chopper lacked the detailed knowledge of the area. The ambushers knew all the back alleys, and what corridors connected to them, and what rooms with a commanding view of the street adjoined those corridors. The gunmen lived there. While the U.S. troops had to move and expose themselves, the Somali snipers could stay hidden until they could get a clear shot at a U.S. target and then fade away into the wrecked cityscape.

Time wore on. A mine blasted one of the Ranger Hummvees and yet another Blackhawk crashed elsewhere. The military command was able to insert a pair of Delta Force snipers into the area where the first chopper crashed—an effort to keep the Somali gunmen from attacking the crew. The men, who fast-roped their way down to the wreckage, were themselves killed before too long.

Meanwhile, the Rangers fought their way to within 200 yards of the first downed helicopter; they were unable to go further because of the intensity of the sniper fire. With dismay, the would-be rescue team found that it had maneuvered itself into trouble. It had managed to split itself in half—one part of the unit covering on one side of a wide thoroughfare and the other part pinned down on the other. Neither could reinforce the other—and each needed help as casualties mounted and ammunition supplies ran low.

As the situation near the city center got worse, a unit of the 10th Mountain Division was called in to back up the beleaguered Rangers. C Company was ordered to make up a rescue convoy. About a dozen Hummvees and five-ton trucks were planned for the force. There was no U.S. armor available—only virtually unarmored vehicles—to make the dash through the streets. Just as telling, the heaviest weapons available to the convoy were 40mm grenade launchers carried by the troopers.

Just prior to 6 P.M., the rescue column was ordered to launch toward the trapped Rangers. Traveling at about 30 MPH, the C Company convoy moved down a wide street, headed for the airport. Within minutes it was being hit by AK-47 fire, fire that came from hidden gunmen manning ambush positions on both sides of the highway. The convoy pushed on for several hundred yards before the trucks rolled to a stop and the troops detrucked. A half-hour firefight followed.

The commander of the operation, concluding that armored vehicles would be needed to reach the Rangers, ordered the troops to withdraw. C company managed to get back to the trucks and drive away—spurred on by an ever-increasing rain of bullets.

With the rescue column from Charlie Company licking its wounds, the commanders called in Alpha Company of the 2nd Battalion, 14th Infantry, which moved to the airport in trucks and Hummvees and then rolled to a staging area near the port area of Mogadishu. There they found about two dozen Russian-made armored personnel carriers (APCs), manned by Malaysian peace-keepers, and a Pakistani unit that was equipped with lightly armored vehicles. The men of Alpha Company got onto the APCs, and the Malaysians drove toward the trapped Rangers.

Somali snipers fired at the armored cars. So did Somali marksmen wielding rocket-propelled grenades. The weapons that had been decisive against men in unarmored trucks could not touch the APCs. In fact, when the convoy was about a third of a mile away from the beleaguered Rangers, the inexorable movement seemed to have dispirited the Somalis; the firing died down. At this point, the troops were ordered to leave the APCs and make the final push afoot. Suddenly, the battle was once again to the liking of the ambushers. As the troopers alighted, the Somali fire re-erupted.

The Americans from Alpha company, caught between Scilla and Charybides, the Somali snipers and the Rangers' perimeter, were nonetheless able to fight their way through and link up. For about two hours, the Rangers and Alpha Company strengthened the perimeter and prepared to fight their way out, calling in helicopter gunships to silence sniper nests. The withdrawal began about 3 A.M.; by 4 A.M. the wounded and the rescuing troopers had retreated out of the maelstrom.

This is a textbook case. Just about everything that could go wrong on an operation did on this one. But the most telling mistake was one that no soldier can ever deal with; it's the kind of misguided policy pronouncement that no protected person can ever resolve.

Political and public relations considerations cause politicians and governments to set up guidelines for action. And when politicians and governments set up stupid guidelines, they end up setting up the conditions for the ambush of their own forces. In this case the U.S. forces lacked the armor they needed to carry them around the urban battlefield. They were bereft of the vehicles they needed to muscle their way out of ambushes.

They lacked that armor because of a political decision made by the U.S. Secretary of Defense, Les Aspin. Aspin was a man with a history of military-baiting who was himself brought into the office to slash back the spending and the power of the military. Despite the long record of violence by the Somali gunmen and the clearly demonstrated need for armor to defend against the ambushers, Aspin had denied it to the troops. He reportedly feared the rest of the world might not look favorably on the United States if tanks took part in what was supposed to be a mercy mission. But he didn't have to, would never have to, crouch in the darkness looking for a slight movement that might betray another man sighting a rifle at his forehead.

There is nothing any military commander can do to break an ambush when bushwhacked by the political hierarchy. That is the first, and perhaps the most important, lesson to be learned from this ambush.

But there are other lessons. For instance, you can't make a bad plan good by wishing it would work. This plan stunk. Even if the Blackhawks had not crashed, the fact that a road convoy was supposed to make its way into the center of "unfriendly" territory and extricate prisoners and their captors was an open invitation to ambush. This operation had all the elements of another disastrous battle plan executed by the British army near Boston on April 19, 1775. It turned out about the same—troops could go forward for a while unhindered, but they couldn't get back because of a host of hasty ambushes set up along their line of march. Screwup number two!

The troops lacked the type of weapons and munitions needed to stage the type of firestorm counterattack that breaks ambushes—screwup number three.

Next came a fundamental flaw—a complete misunderstanding of the roles and capabilities of special units. Rangers, Delta, Special Forces, SEALs, Marine Force Recon, and similar units rely on stealth and speed to carry out their missions. Their role is to get in, act, and get out. They do it fast—and clandestinely. Special units aren't taught how to hold ground—they don't have the weapons to hold ground, and their mission role is violated when they are used as regular infantry. The decision to move some of the Rangers to the scene of the Blackhawk crash, set up a perimeter, and hold ground, all of these violated the basic concept of special warfare units. Chalk up another contributing factor to the debacle.

There was no planning for the helicopter crew rescue operation, unless an ad hoc, seat-of-the pants decision can be called planning. It is hardly any wonder that the Rangers ended up with a split element—one of the most basic mistakes any tactician can make—and could not reach the downed aircraft.

Sending lightly armed troops in unarmored vehicles along a road made to order for ambushers in order to extricate the Rangers never made sense either. The Rangers themselves were lightly armed, had access only to unarmored vehicles, and were stuck in the same ambushers' terrain. It's hard to imagine how anyone could think things would be better for C Company than the Rangers. That decision was the capstone to this monumental "invitation to an ambush." When adding in other things, like ordering Alpha Company troops out of the protection of their armored vehicles, it's a miracle anyone survived this series of multiple ambushes.

U.S. and allied operations in "Mogadishu 1993" constitute a casebook on what not to do when breaking up an ambush. Fortunately, there are far better examples of what should be done.

PHUOC LONG AMBUSH (VIETNAM, 1960s)

A company of the 37th Ranger Battalion, engaged in a road-clearing operation in Phuoc Long Province, moved from its camp in the morning. The VC, estimating correctly that the unit would return over the same route later in the day, established an ambush along the edge of the jungle overlooking the road.

The Ranger company did, in fact, return over the same road. However it was deployed with two platoons abreast, each moving 150 yards off the sides of the road. The company headquarters and the other platoon followed in trace behind the right-hand platoon. At 2 P.M. the lead platoon on the right discovered the left flank of the ambush. The platoon sergeant immediately deployed his troops and began firing at the VC. The company commander ordered the trailing platoon to swing to the right and assault the ambush while the other two platoons and the mortar section supported by fire. After 30 minutes, the VC retreated. A total of seven VC were killed and three carbines captured. There were no friendly casualties.

In this case the company commander was prepared for an ambush. He had dispersed his unit on the march, avoided the road, and provided for an adequate reserve. His quick response and sound tactics, coupled with the skills of his troopers, turned a possible VC victory into defeat.

Units must be prepared for an ambush at any time, especially while moving. When they're not ready, the result is disaster.

NAGA TRIBESMEN AMBUSH (EASTERN INDIA, JUNE 30, 1993)

Tribal rebels fighting for independence in eastern India ambushed an army convoy on June 30, 1993. They shot and killed 26 soldiers in the rebels' largest ever offensive. According to The Press Trust of India news agency, official sources said that 10 others, including three members of the separatist tribal National Socialist Council of Nagaland (NSCN), were killed in the rebel ambush on a road linking Manipur with neighboring

Nagaland state. At least 20 people, including civilians, were wounded in the crossfire. Twenty other soldiers were injured, many of them seriously.

About 120 troops were traveling in trucks on a winding mountain road in Manipur state near the Burmese border when they were attacked. About 40 attackers, positioned on both sides of the tree-lined road, fired automatic guns at the unprotected trucks. One truck containing the troops' ammunition exploded. It was the most serious attack on security forces by the guerrillas, who had been waging an off-and-on campaign for four decades. Guerrillas from the group killed 21 soldiers when they attacked an army convoy in 1982 in the same region.

The trucks were unarmored and unprotected. When the ammunition exploded, it was all over but the shouting—and the crying and the dying. Ammunition supplies and explosives need special handling. Secondary explosions are more than demoralizing for troops caught in an ambush. They produce casualties of their own and indicate that those caught in the ambush may soon be totally defenseless because of a lack of ammo. Part of the preparation for an ambush is making certain that troops have adequate quantities of munitions in order to stage the violent counterattack that breaks up the bushwhacking.

The ability to return fire, even under the most difficult of conditions, can be crucial to surviving an ambush.

AMBUSH OF POLICE (SOWETO, SOUTH AFRICA, MAY 5, 1993)

Gunmen killed four police officers in South Africa's Soweto township in an ambush that came close to wiping out many more. Five officers were wounded in the May 5, 1993, attack on a truck that was carrying 23 police officers home from the night shift. The officers had just completed their shift at Dobsonville police station and were on their way home when their vehicle was attacked from two sides by gunmen armed with AK-47s and other weapons.

Although trapped in their bullet-riddled vehicle by the tangle of dead and wounded, several of the uninjured officers in the truck managed to return fire, driving off the attackers. There was no immediate claim of responsibility, but the Azanian People's Liberation Army (APLA), armed wing of the radical Pan Africanist Congress (PAC), later claimed responsibility; it had publicly stated that police were high on its list of targets.

The APLA has a history of ambushing, but in the main it is a history of striking at the defenseless. The ability of the ambushed policemen to shoot back, even though technically outgunned and caught by surprise, forced the ambushers to break off the engagement.

But sometimes there is no chance to return fire—and that is particularly true with bomb and mine attacks.

IRA AMBUSH OF ULSTER DEFENSE REGIMENT (DOWNPATRICK, N. IRELAND, APRIL 9, 1990)

Four members of the Ulster Defense Regiment (UDR) died when a command-detonated land mine blasted their armored Land Rover near Downpatrick, County Down, on April 9, 1990. The vehicle was one of a pair patrolling a country road when it fell into the explosive ambush. The attackers allowed the first vehicle through the killing zone and then blasted the follow car. The explosion hurled the hit Land Rover across a hedge, tossing it nearly 100 feet into a field. The explosive—which had been left in a culvert—created a crater 15 feet deep, 50 feet long, and 40 feet wide. Four UDR members in the first vehicle and two civilians in other cars were hospitalized as a result of the concussion from the blast. The IRA's South Down Brigade accepted responsibility for the attack, the most deadly IRA action against the citizen-soldier group in seven years.

Command-detonated roadside bombs and bombs placed in culverts, a technique mastered by the IRA and a variety of Middle Eastern groups, are nearly impossible to defend against. There is no indication, short of searching every culvert or planting remote sensors in every culvert to warn when someone approached one, that this ambush could have been avoided.

Clearly, good intelligence on the part of the IRA, knowledge that the patrol's route would take the Land Rovers over the culvert, and the failure to intercept the movement of the explosive before it was emplaced contributed to the success of the attack. The follow vehicle may have been chosen as a target because the ambushers were timing the first vehicle as it crossed the culvert.

Ambushers learn quickly what works and what doesn't. And sometimes lightning does strike twice.

HEZBOLLAH ATTACK (SOUTH LEBANON, OCTOBER 25, 1992)

On October 25, 1992, Hezbollah militants exploited a chink in Israel's armor, setting off a remote-controlled bomb in south Lebanon that killed Israeli soldiers; five others were wounded, three critically. The bomb went off on the southern edge of the Bekaa Valley inside Israel's self-declared security zone in south Lebanon. It was the deadliest attack by pro-Iranian Hezbollah guerrillas on Israeli forces in the south since a Hezbollah suicide bomber killed seven soldiers in October 1988.

The Israelis had sent an "early bird" patrol along the Ahmadiyeh Road as usual at about 8 A.M. The road, built by Israel years before, was used exclusively by the Israeli Defense Forces (IDF). The patrol, which was making a security check before the passage of a regular weekend supply convoy bringing water, food, and troops to IDF positions, checked only for bombs planted under or at the side of the road. The bomb was planted in an embankment. It went undetected, and when it exploded, it blew high. The blast was at just the right height to inflict maximum casualties on a truck. The ambush of the supply convoy was identical in technique to an attack in the south in October 29, 1991, which killed three Israeli soldiers. In each case the bomb was planted in an embankment above road level to cause maximum casualties.

In this ambush it is significant that the road was used only by Israelis. Troops had forgotton to look up and were too busy looking down and to the side. They had also forgotten how well a similar attack less than a year before had succeeded. Those were ingredients in the recipe for disaster.

There was another factor that may have contributed to the success as well—the victims may have been counting on the fact that they were only 120 yards from a UN post in an area under Norwegian supervision. Apparently what they didn't know did hurt them. A UN spokesman said the post was unoccupied when the bomb exploded.

Hezbollah obviously had good intelligence. In this case, they must have been watching the road from a distance because, when the ambushers detonated the bomb by radio, the timing was perfect. Detection of the detonation team might have changed the situation.

There are a number of ambush types using bombs. Sometimes the targets are civilians.

ELN AMBUSH (BOGOTA, COLOMBIA, SEPTEMBER 28, 1988)

Colombia's so-called National Liberation Army, the ELN, carried out an ambush bombing in Bogota against a U.S. oil company executive on September 28, 1988.

In the attack, a remote-control explosive device exploded, damaging an armored car carrying the manager of Texaco operations as he drove from work to home. The bomb, concealed in a cart parked along the curb of the road, was command-detonated from a safe distance. It had been positioned so as to concentrate the explosive force on the passing vehicle, but the blast apparently was set off a fraction of a second early. Even at that, the force was sufficient to disable the vehicle. A guard and a passerby were wounded.

In the claim made later that day, the ELN said the attack was designed to protest the company's exploitation of Colombian resources and financing of paramilitary organizations. Both the attack style and the reasons given were typical of the ELN.

The group's history shows it is capable of a wide range of tactics and has a good weapons selection. In addition to explosives, the group uses small arms, semiautomatic weapons, grenades, and antitank rockets. An important lesson to be learned from this attack is that while it is impossible to prevent a remotely controlled bomb ambush along a roadway—after all, who can have every cart and parked car checked along the route before driving home?—such bombs have to be timed exactly. Speeding up and slowing down, accelerating slightly at some corners, perhaps slowing and speeding up by as little as five miles an hour over straight-away stretches, can make an ambusher's job much more complicated. It makes driving a lot more complicated as well.

CINCHONEROS AMBUSH (TEGUCIGALPA, HONDURAS, JANUARY 25, 1989)

Gustavo Alvarez Martinez, who headed counterinsurgency operations as the Honduran military commander-in-chief from 1982 to 1984, was killed, along with his driver, in an ambush claimed by the Cinchonero movement on January 25, 1989.

His vehicle was intercepted by rebels, disguised as employees of the national telephone service, shortly after he left home in the morning. The attackers shot up his vehicle about three blocks from his home when it halted at a stop sign. Blasts from Uzis blew out the tires and disabled the vehicle. Then rebels poured automatic weapons fire through the windshield. Alvarez Martinez, who had undergone a religious conversion and relied on the power of the Bible for his defense, was unarmed. He had refused bodyguards offered by the government.

There are obvious lessons to be learned—not the least of them being to take reasonable precautions. Some felt Alvarez Martinez had a desire to be martyred; his more unkind critics claimed this desire stemmed from his past attacks on rebels. His refusal to take the most elementary precautions, from having a bodyguard to carrying any type of defensive weapon, cost him his life and the life of his employee as well.

The disguise of the attackers and the fact that the attack occurred at a stop sign three blocks from his home illustrate the truism that a key to successful ambushes is getting the target stopped—whether the target is halted by gunfire or a stop sign doesn't matter. It also showed he had been under observation; observant bodyguards, if Alvarez Martinez had had them, might well have spotted the terrorist snoopers and saved the lives of two people.

The success of the attack also shows that small groups are as dangerous as a standing army. The Cinchoneros, also known by the initials MPL for its Spanish name, was believed to number fewer than 200 at the time. But size doesn't mean anything to a group intent on carrying out an ambush, particularly a group with experience. Experience was something the Cinchoneros had. It was in San Pedro Sula, for instance, on July 17, 1988, that the MPL attacked U.S. military men as they left a discotheque. The servicemen were shot at and targeted with an explosive device; some were pursued as they fled the determined attackers. Five U.S. troopers were wounded in that ambush.

But even experience, however, doesn't always guarantee success when it comes to ambushes.

"REVOLUTIONARY ORGANIZATION NOVEMBER 17" (GREECE, APRIL 24, 1987)

The Greek terrorist group "Revolutionary Organization November 17" is considered one of the best organized and efficient forces in the world. But when it ambushed a bus carrying U.S. and Greek military personnel, it had only limited success.

In this attack the group attempted a mass attack, not the group's forte. Their target was a bus carrying 35 U.S. and Greek military personnel, as well as American dependents. The bus was a shuttle that regularly transported personnel between air bases. Obtaining information on the operation and scheduling must have been simple . . . a few observations would have been all that was required.

The idea of the attack was simple. A powerful explosive was placed along the road at a spot the bus routinely passed on its rounds. About 300 feet of detonating cord was strung between the ambush site and a hut where the attackers hid. The ambush spot was just past railroad tracks, a semibarrier in the road that had forced the bus to slow. That gave the attackers a better opportunity to time the blast for the moment when the bus would be in the killing zone. They apparently used a tree alongside the road as a line-of-sight marker, one by which they gauged the proper moment to detonate the explosive. As it turned out, the attackers either jumped the gun or misjudged the effective radius of their explosives; the bomb went off "prematurely."

Since the bomb was planted adjacent to a roadside retaining wall, the main blast effects were directed toward the road and the bus. Though the bomb prematured, the blast was severe enough that the driver lost control of the bus. The vehicle continued out of control for about 25 yards before hitting the retaining wall next to the road. Eighteen people were injured in this attack, but the single serious injury occurred when a U.S. sergeant was thrown through the windshield of the bus. A car driven by a Greek national was traveling parallel to the bus at the time. That car was spattered with shrapnel from the blast and the driver suffered slight injuries. The attackers escaped.

The failure of this attack underlines a previously discussed issue—the difficulty of timing the blast to catch a moving object. The terrorists made every effort to solve that problem by hitting the bus at a railroad crossing, where it had slowed.

It's worth noting that November 17, with only 20 to 25 members, is a past master at the use of ambushes. It is known for its highly efficient and generally deadly assassination attacks on individuals. It has generally selected solitary targets, stalked them, and killed them. This Marxist-oriented group started its effective ambush operations with the assassination of U.S. Embassy officer Richard Welch and has used the same pistol in subsequent assassinations.

That bus attack was consonant with November 17's specialty: ambush-type assassinations and attacks by teams of up to three. November 17 victims are often attacked near their homes or offices, eliminating from the security equation any prophylactic measure of taking differing routes to and from work. But this ambush differed in one respect—most November 17 attacks are directed at carefully selected individuals who often have, or should have, the training and security screen that would keep them alive. The people on the bus were not selected to be targets as individuals—most were just Americans and they were just in the wrong place at the wrong time. Sometimes there is nothing that can be done about that.

NEW PEOPLE'S ARMY AMBUSHES (ANGELES CITY, PHILIPPINES, OCTOBER 28, 1987)

Ambushes against targets of opportunity are particularly hard to counter. Communist "sparrow" hit-squads carried out a series of four coordinated attacks near Angeles City on October 28, 1987. All occurred within a 15-minute time span. Three Americans and a Filipino bystander were killed; one American escaped unscathed in an attack on his car.

In the lead-off attack, a uniformed U.S. Air Force sergeant had gotten off a jeepney transport near his home and was walking away when a trio of Filipino males overtook him from the rear. At about 5-feet's distance, one of the men opened fire with a handgun, hitting the doomed noncommisioned officer (NCO). When he collapsed on the ground, the three gathered around him and fired at him repeatedly. One of the assailants then shot him through the throat before all three got into a waiting jeepney and escaped.

Contemporaneously, a four-member attack team waited under a tree by the side of a road off the main highway but near the main gate of Clark Air Force Base. A retired USAF NCO, a Philippine-born U.S. citizen, was driving by when the attackers hailed him and approached the car. Spreading out in a semicircle they opened fire at the driver with .45 handguns; five of the 13 rounds struck and killed the driver. The hit team then calmly walked back in the direction from which the victim had come, heading for a waiting jeepney that was their escape vehicle. In doing so they passed another car that had driven up to the ambush scene. One of the attackers stared at the witness as if deciding whether to kill him as well, but all continued on and escaped in the waiting vehicle.

At about the same time, an airman who slowed his car to obey a stop sign at an intersection was suddenly fired on by a hit-squad. There were from four to six attackers waiting by the side of the road. The mortally wounded driver lost control of the vehicle, which continued ahead and struck a utility pole across the street from the ambush spot. The attackers moved there and pumped more shots into the car and the airman. While they were doing so, another vehicle approached the site from the side. Thinking there had been an accident, the driver slowed down. One of the attackers then turned and fatally shot the driver of the second car, a Filipino businessman. The witness' foot came off the brake, and the car lurched forward and to the right, crashing. Those attackers escaped in a jeepney and a three-wheel cycle that had been parked by the side of the road.

In the fourth attack, a U.S. Air Force officer slowed his car at a T- intersection to make a turn. As he did so, he noticed one man pulling a .45. The officer immediately accelerated and began turning. At least two gunmen, and possibly a third, fired at the car and hit the driver in the chest with one round. However, the shot was deflected by a checkbook kept in the pocket of his flight jacket, and the car sped through the killing zone.

The randomness of the ambusher's bullet is quite apparent in this case, as is the variety of methods used to slow or stop the victims so that they could be killed. What also stands out is that the one victim who survived was observant enough to see the gun being drawn and quick enough to take evasive action. Being observant is important, but sometimes observations have to be made long before the event if they are to be effective.

HERRHAUSEN ATTACK (BAD HOMBURG, GERMANY, NOVEMBER 30, 1989)

On November 30, 1989, the Red Army Faction (RAF), considered by security specialists at the time to be semi-dormant, staged a spectacular explosive ambush and killed the head of Deutsche Bank. Alfred Herrhausen was riding to work in an armored limousine when the car was blasted in Bad Homburg, the suburb of Frankfurt where he lived. The driver of Herrhausen's car was severely injured in the 8:30 A.M. attack, which occurred less than a mile from the banker's home. The powerful remote-control bomb caught the banker's car, blowing it into the air; forcing open the trunk, hood, and doors; and setting the vehicle aflame. A lead car and follow car carrying bodyguards were unaffected by the blast. The RAF, which had not carried out a successful major attack in three years, claimed responsibility in a letter left at the scene. The letter, bearing the seal of the group, was signed with the name of the Wolfgang Beer Commando, RAF.

The explosives were concealed beneath a bicycle left near the roadway. They were triggered by a sophisticated device and were activated by a wire that had been run into a park near the road. In some places the wire had been strung across a macadam sidewalk by chiseling the asphalt out, laying the wire in place, recovering the wire with asphalt, and tinting the work to disguise the fact that anything had been done.

In fact, it was the second time the cable had been emplaced.

A cable that had been laid in the area was found by an employee of a nearby spa the month previous to the attack and had been removed—but the attackers replaced it! No one connected the mysterious cable to anything nefarious until after the bombing.

A man in a jogging suit, wearing earphones, was seen in the park and was believed to have activated a light-beam device on the approach of the banker's vehicle. When the banker's car broke the beam, the explosives detonated. The force of the blast was directed toward the right rear of the vehicle, where Herrhausen was seated.

A car believed to have been used by two men in escaping the area was later found in another Frankfurt suburb, Bonames. The escape car had been rented October 17, suggesting the length of time that the attackers had been actively engaged in the final preparations.

Could this ambush have been broken? In theory, yes. Practically, no. Security teams don't routinely check parks to see whether someone runs a cable under a walkway. Park personnel might have had reason to wonder what was going on when they found the cable and work on the walkway, but they had no reason to connect a simple piece of wire with Herrhausen, the RAF, or a bomb.

The timing of the explosion, always a delicate matter in a Lebanese car bomb attack, was probably the most crucial factor to the success of the attack. The attackers effectively canceled timing from the equation when they used a light-beam triggering device. Of course, had Herrhausen been seated elsewhere in the car things might have been different. But the attackers knew where he usually sat.

This attack stands out as a classic in terms of planning and execution. This was a case where simply being observant would not have changed the outcome much. But that is not always the case. In most instances being observant and taking action is the key to surviving an ambush. That can hardly be sufficiently stressed.

ALDO MORO AMBUSH (ITALY, MARCH 16, 1978)

Aldo Moro knew he was a target. He was so clearly a target that he rated a whole security detail. On the morning of March 16, 1978, he left for work but never arrived.

His entourage was in two cars. A driver and a bodyguard accompanied him in the lead Fiat 130. Three bodyguards were in a follow car. Just before an intersection controlled by a stop sign the driver of a small white car passed Moro's cars, then jammed on his brakes at the intersection. Moro's car hit the white car; the follow car slammed into Moro's vehicle.

Two men quickly got out of the white vehicle, pretending to be ready to inspect the damage. Approaching Moro's vehicle they whipped out weapons and killed both the driver and bodyguard. Meanwhile, four men standing nearby, dressed in airline uniforms, opened fire on the follow car and the guards.

Moro was forced out of his car, hustled into a waiting vehicle, and whisked away. The entire attack—a prisoner ambush—was over in about 30 seconds. The ambush succeeded for a number of reasons.

First was surprise. None of the guards was alert enough or suspicious enough to draw a weapon, even in light of the fact that a car whipped by to pass, slammed on its brakes, and caused a crash. No one was suspicious, despite the threats against Moro. No one was suspicious that it was morning and Moro was on his way to work—the most significant marker of an ambush.

The second reason for the success was just plain inaction, surprise or not. The driver and bodyguard allowed the ambushers to walk up to them without ever stirring in the car, without trying to move the vehicle or get out. They froze completely.

The third factor in the Moro ambush-kidnap—and his eventual murder—was intelligence. The terrorists had the ability to predict where Moro would be and the time he would be there. Although Moro's security force had five possible routes to take to and from the office, he told people in advance which one he was taking!

Stupid mistakes are easy to make and difficult to understand later. Someone always believes "it can't happen to me."

RENAMO TRAIN AMBUSH (RESSANO GARCIA, MOZAMBIQUE, FEBRUARY 14, 1990)

On February 14, 1990, RENAMO rebel forces used a remote-control mine to derail a passenger train near

Ressano Garcia. After six cars in the unescorted train rolled off the tracks, the attackers opened fire on the wreckage with automatic weapons. There were about 100 casualties, including 55 slain. Some of those who survived the ambush were kidnapped by the rebels.

In this classic train ambush, RENAMO was at a place they weren't expected to be—and worse, there were no effective security plans in place. The mine damaged the rail bed and derailed the train—just as it was intended to do. The rebels, without any resistance, were able to kill and kidnap anyone at will.

The "it can't happen to me" or "lightning won't strike twice" attitude is all too common. It is also deadly.

TRAIN AMBUSH (KAMPOT, CAMBODIA, AUGUST 2, 1993)

On August 2, 1993, Cambodian attackers laid mines on a railroad track and raked a train with gunfire and rockets, killing at least 10 people and injuring 30. The afternoon ambush took place about 14 miles east of the provincial town of Kampot in the south of the country. The attackers first exploded mines on the track of the Phnom Penh-bound train, then raked the cars with small arms fire and shoulder-launched rockets. Units from the U.N.'s French paratroop battalion based in Kampot province reached the ambush scene about 15 minutes after the attack. The ambush was in the "violent triangle," known as a sanctuary for Khmer Rouge guerrillas who had attacked trains in the region before. In fact, on July 25, gunmen believed to belong to the group attacked the same Phnom Penh-bound train about 25 miles northeast of Kampot, injuring several passengers. And the previous May rebels attacked the same line in the same general area, derailing the train.

The "lightning won't strike twice" theory is bunk. If anything, a successful ambush is likely to be repeated.

Railroad officials aren't the only ones inclined to miscalculate. Rebels sometimes suffer from an "it can't happen to me" syndrome. And it does happen to them.

SAS AMBUSH OF IRA TEAM (COAGH, NORTHERN IRELAND, JUNE 3, 1991)

Three members of the IRA on "active service" were killed by commandos of the Special Air Services (SAS) in an ambush on June 3, 1991. The undercover troops ambushed the three—Tony Doris, Lawrence McNally, and Peter Ryan—in their car at Coagh. The trio were in a hijacked car and were planning to attack some unidentified target when soldiers concealed along the town's main thoroughfare lashed the vehicle with automatic weapons fire in the breakfast-time ambush. The car caught fire after crashing. Two of those inside struggled out, their clothes aflame, to die in the street. The British military, in its statement, only acknowledged that a "special covert team" was acting on intelligence information.

The British military knew an IRA attack was laid on. It knew who the attackers were, how they were going to get to their target, when they were going to do it, and what they were driving. An informant had apparently talked—something that underlines the need for good security and the penalty that will paid when security fails. Strict enforcement of "need to know" rules and reliability of those you're working with are important no matter which side of the fence you're on. The IRA men just could not believe it could happen to them.

The British have had notable success in laying ambushes against the IRA. For instance, in May 1987, an IRA attack team of eight people was ambushed and killed by British commando forces at Loughgall, Northern Ireland. The active service rebels attacked the police station in the village with a bomb and a mechanical digger, but it was wiped out in an ambush by undercover commandos.

It is also important to note that the SAS and commandos aren't saddled with a gentlemanly view that the enemy should have a chance to either surrender or fire back. There was no "code of the West" gunfight on a dusty street. It was a simple and effective bushwhacking. The troops were true to the ambusher's credo: Shoot first. Shoot fast. Shoot last. Shoot to kill.

Sometimes, when that becomes too apparent, the government gets embarrassed. In January 1993, a Northern Ireland inquest—which generated considerable news coverage—ruled that undercover British soldiers shot and killed one of the IRA's most feared guerrillas without giving him a chance to surrender. Seamus McElwaine had been killed and a second IRA guerrilla wounded in 1986 near the village of Roslea, close to the Irish border, as they prepared a bomb ambush against security forces.

McElwaine was on the run after escaping from the top-security Maze prison with other members of the IRA. He was serving a life sentence for the murder of two members of the security forces. The coroner's court jury in Enniskillen said that troops had not challenged McElwaine before he was hit by an initial burst of gunfire. The jury also said they fired again five minutes later as McElwaine lay injured. The jury chose not to believe four soldiers involved in the incident who insisted the two men were challenged just before they opened fire. Instead, they found the word of wounded guerrilla Kevin Lynch more plausible. He told the inquest that McElwaine was shot after being questioned by soldiers as he lay wounded.

The soldiers' and government's case in the court of public opinion wasn't helped, either, by the fact that the Defense Ministry issued a public-interest immunity certificate preventing disclosure of some details relating to the army's undercover operation.

The jury's findings in the case made it possible for McElwaine's family to take legal action against the Ministry of Defense.

SOLDIERS' AMBUSH OF STONE-THROWERS (BETHLEHEM, ISRAEL, AUGUST 19, 1989)

Security forces posing as tourists loitered near a popular Christian pilgrimage site until Arab youths carried out habitual taunting and stone-throwing attacks against soldiers in Bethlehem. The "tourists" then drew weapons and fired at the stone-throwers, killing one and wounding three in this early morning ambush on August 19, 1989.

This was another case of a security team ambush against militants—but one where no informant was needed to catch the insurgents unawares. The youths had a habit of making the stone-throwing attacks—a type of open-air ambush in itself. They had established a pattern that was detectable as to time and place. Unsophisticated in comparison with the Israeli security forces, the youths failed to provide for adequate security during their own attacks and neglected to use maximum cover and concealment.

The tactic of putting disguised security forces on the streets, posing as everything from militants to women, had worked time and again. It was one thing to describe the process, however, and another to see how it actually worked. Later, on June 21, 1991, film clips aired by Israeli television showed exactly how military personnel used a variety of disguises to move undetected through Arab areas and ambush Palestinian militants. The ploy was admitted—and praised—by army spokesman Brig. Gen. Nachman Shai who said the undercover troops were responsible for apprehending "hundreds of wanted individuals." The showing of the film clips and the government's admissions created a community controversy. The "ungentlemanly" deception became fodder for Palestinians and damaged the government's positions, even among some Israelis.

When such deceptions are in use, insurgents begin to suspect everyone. All tourists, or news reporters, or whatever the cover of the week is, become subject to attack. Generally the genuine object is all too easy to hit, too. Tourists, even those with the training that should allow them to deal with ambushes, seem to leave their caution at the airport or bus station.

MPLF ATTACK OF BUS (TEGUCIGALPA, HONDURAS, MARCH 31, 1990)

Attackers struck a bus carrying U.S. servicemen, injuring eight soldiers, two of them seriously, in this March 31, 1990, ambush. A trio of gunmen shot up a bus filled with U.S. Air Force personnel, covering it with a curtain of automatic weapons fire. The driver kept the bus moving and ran through the ambush zone.

The attack occurred about six miles north of the capital city of Tegucigalpa as the bus was ferrying 28 airmen from a recreational trip to the beaches of Tela back to the Cano Soto Airbase. The attack was claimed by the Morazanista Patriotic Liberation Front (MPLF) in a call to a radio station.

The soldiers, returning from merrymaking at the beach, with someone else driving, were hardly looking for gunmen to spray their bus with lead. They weren't prepared to duck and take cover, much less to fight back. The key to their survival was in the hands of the driver, who was able to keep the bus rolling through the killing zone. The driver, in this case, made the difference between injuries and death.

ATTACK ON BUS OF ISRAELI TOURISTS (ISMAILIYA, EGYPT, FEBRUARY 4, 1990)

A pair of masked attackers shot up an Egyptian-owned bus filled with Israeli tourists in an ambush along a main desert road. Nine people died as the result of the February 4, 1990, attack. Some 17 Israelis were wounded.

The attack took place between Ismailiya and 10th of Ramadan City after a security escort who had accompanied the bus from Rafah left the bus at Ismailiya. That was part of a normal routine.

The attackers, who were using a rented car, then reportedly cut the bus off and forced it to a stop. Alighting from the car, they ordered the driver, a Palestinian, and the tour guides off the vehicle, before lacing the Rafah-to-Cairo bus with automatic-weapons fire. Only half of the four grenades they tossed at their unarmed targets exploded, limiting the casualties. Yet some who attempted to escape were cut down in their tracks.

The attack was claimed in the name of an unknown group, Organization for the Defense of the Oppressed in Egyptian Prisons. The caller said the attack was retaliation for the Egyptian arrest of Muslim militants and alleged torture.

Before militant fundamentalism was recognized as a major threat in Egypt, there were nonetheless concerns about the safety of Israeli visitors. That's why there was a security escort accompanying the bus for part of the trip. The removal of the security escort at Ismailiya meant that any attack would be staged thereafter. A security escort that doesn't go the entire way is a setup to ambushers.

These attackers used the threat of collision to stop the bus—though the bus driver could have thwarted the attack simply by using his vehicle as a battering ram. A properly trained driver, alert to the fact that the passengers he was carrying were likely targets for a terrorist attack, would have been able to damage the attackers' car so badly that they might have even been unable to escape. He might even have killed some of the ambushers.

The ambushers were unusually careful about who they killed. They had the necessary time to cull out targets from nontargets. Those people on the bus, many of whom would have been more aware at home and would probably have been armed, were lulled into a false sense of security because they were on a trip—it couldn't happen then or to them. Whether they, unarmed, could have counterattacked the gunmen themselves is problematical. But in any event, those who were cut down trying to escape did themselves no favors. The most successful were those who took cover and sought concealment when pinned down in the killing zone. Only the failure of half the grenades, and the fact that the ambushers broke off the attack before making certain everyone had been killed, kept the attack from being more successful than it was.

The diplomatic repercussions of this attack were felt for months, demonstrating how a single ambush affecting the lives of a few dozen people can have serious effects on entire nations and peoples.

The Israelis nearly always seem besieged—and diplomatic repercussions come from other kinds of ambushes.

ISLAMIC JIHAD BUS AMBUSH (EIN NETAFIM, EGYPT, NOVEMBER 25, 1990)

A man dressed in an Egyptian border guard's uniform sprayed a civilian bus and three Israeli military vehicles with automatic weapons fire, killing four people and wounding about two dozen on November 25, 1990. The gunman then fled back into Egypt following the 7 A.M. attack near Ein Netafim, northwest of Eilat, in an area where there were no major impediments to a border crossing.

Three soldiers and a driver for the national bus company were killed in the attack; 23 civilian defense workers who were bus passengers were wounded.

The single gunman took up an ambush position in a roadside ditch; from there, he could fire on vehicles as they passed. The driver of an army van was hit first, then a second military vehicle came by and was punctured by rounds. The driver of an empty military bus, thinking an accident had taken place because of the position of the other two vehicles, stopped and got out. He was shot and killed at close range by the gunman. A civilian bus followed. The gunman killed the driver, then circled the stopped vehicle, which was carrying civilians employed at military installations in the area. He sprayed it and the passengers with dozens of rounds. A civilian guard shot and wounded the gunman, who fled back into Egypt just as army troops arrived on the scene. The Egyptian border policeman assigned to the area was later arrested by Egyptian authorities. The Islamic Jihad Movement praised the attack and claimed one of its "units" was responsible.

As well-trained and alert as Israelis are, the simple fact is that all of the vehicle drivers and many others failed to recognize an ambush was taking place until it was too late. Some of the drivers willingly stopped in the killing zone, thinking themselves Good Samaritans.

The attacker consistently targeted the drivers of the vehicles as a priority—making escape from the killing zone that much more difficult.

The counterattack by a civilian guard aboard the bus halted the attack—coupled, of course, with the imminent arrival of a superior force of Israeli army troops. The gunman, who clearly had planned his route of retreat carefully, fled successfully back into Egypt using the international border as a shield.

Well-planned ambushes allow time to carry out the attack and still escape. A border is hardly a necessity. Rugged terrain and poor communications that give the attacker a head start can serve the same purpose: to hinder pursuit.

KURDISH BUS AMBUSH (BINGOL, TURKEY, MAY 24, 1993)

Kurdish rebels, shattering a two-month-old cease-fire, killed 31 Turkish soldiers and four civilians in a bus ambush on May 24, 1993. The bus carrying the military recruits and their teachers was stopped by PKK rebels, who had set up roadblocks nine miles outside Bingol at dusk. A group of 150 rebels attacked the bus and destroyed five vehicles.

They first stopped the bus, carrying mainly soldiers in civilian clothes to their units. They forced the passengers from the bus and harangued them on the Kurdish cause before the shooting began. Troops searching the ambush area early the next day found 35 bodies in a ravine near the road and the charred wreckage of the bus and five other burned-out vehicles. Troops and helicopters were assigned to hunt for the attackers and their hostages in the rugged area where the ambush took place.

While the PKK and the army had been fighting a bloody, ongoing war, the PKK leadership had called a unilateral cease-fire months before. It seemed to be holding. The military, which would not have sent unarmed troops and an unescorted bus through the area six months before, got lazy and relied on the truce.

Had the military had better intelligence, it would have known that some PKK commanders, unhappy with the truce, had decided on their own to write "finis" to the stand-down and to do so in blood. Without consulting anyone they carried out the successful—and easy—ambush and slaughter. In the aftermath of the massacre, the government canceled an amnesty that it had just granted that day. Troops resumed full-fledged war against the PKK.

Led to believe that the PKK was still in a stand-down, the driver did not try to force his way through the ambush/road block, which is as understandable as the government's failure to send an escort with the bus in the first place. But hindsight is always best, and caught without weapons or any means of defense, it is clear the victims never had a fighting chance.

Car bombs are among the tactics that leave the ambush target with little way to respond and no fighting chance.

ULSTER CAR BOMB (MOY, NORTHERN IRELAND, FEBRUARY 24, 1993)

A careful off-duty Royal Ulster Constabulary constable,

Reginald Williamson, died on February 24, 1993, near the town of Moy in North Armagh. A bomb, believed to have been attached to the underside of his car by magnets, exploded. Williamson and his girlfriend were driving home in separate cars after an evening out when the bomb exploded. Williamson's brother Freddie, who was a member of the British army's Ulster Defense Regiment, had been killed by the Irish National Liberation Army in 1982 when his car was ambushed by gunmen in the same area.

Williamson had been particularly wary. He was normally careful to check his car for booby traps and to vary his driving routes, but in this case he just wasn't careful enough. The Irish Republican Army claimed responsibility for the attack.

Most Ulster car bomb-type ambushes—where the explosive is attached to the car and is detonated by timer, remote control, or some other device—are preventable. Not all car bombs can be easily, or even possibly, defeated.

BOMB AND PRISONER AMBUSHES (LEBANON, MAY 1989)

The spiritual leader of the Sunni community in Lebanon, Sheik Hassan Khaled, was killed in a May 16, 1989, car-bomb ambush on a crowded Beirut street; 21 others were killed and scores of people were wounded. The stand-off attackers used remote control detonators to set off a 330-pound explosive charge. The bomb exploded as Khaled's security convoy traversed the Aishe Bakkar community between his office and home, where Khaled was returning for lunch. The bomb demolished nearly every structure and vehicle within a 150-foot radius.

That same day a trio of West Germans was taken hostage by Muslim extremists at Sidon. The victims were in a two-car convoy when they were stopped by three masked men with automatic weapons. Two of the kidnapped men, Heinrich Streubig and Thomas

Kemptner, were bundled into the rear seat of a Mercedes; Petra Schnitzler was forced into the trunk. The kidnappers later switched vehicles and took Streubig and Kemptner with them; they freed Schnitzler. The West German government received a demand for the freedom of two brothers, Mohammed and Abbas Hammadi, jailed by the Bonn government in exchange for the release of Kemptner and Streubig.

This was a classic example of the Lebanese car bomb attack that spares no one. A cleric was on his way home for lunch, a daily habit that cost him his life. The assassins knew where and approximately when to kill him; the exact time was left to circumstance. The wide-ranging destruction the car bomb inflicted is typical of this type of ambush and proves that sufficiently dedicated ambushers, with sufficient cunning and no concern for others' lives, can effectively kill any target—even one who has an experienced security cordon.

Security teams, no matter how large and effective, simply are unable to stop all attacks.

ROADSIDE BOMB AMBUSH OF PRESIDENT (BEIRUT, LEBANON, NOVEMBER 22, 1989)

A heavy-duty remote-controlled bomb caught the security convoy of President Rene Moawad in West Beirut. The 1:45 P.M. blast killed Moawad and a dozen of his bodyguards in a 10-vehicle convoy on November 22, 1989. Other high-ranking government officials, including the prime minister, were in another car behind Moawad's, but they escaped uninjured. They were returning from a reception at Sanayeh, marking the 46th anniversary of the country's independence, when they were hit by the explosive ambush. At least 13 bystanders were killed in the searing blast, which went unclaimed. The 400-pound bomb was hidden in a small shop along the street. It was powerful enough to blow the presidential limo from the roadway and twist it into unrecognizable metal parts. The crater went 6 feet deep and extended 30 feet across.

A significant feature of this car bomb attack is that the blast was focused enough to shred Moawad's car, while leaving other cars only damaged and their occupants uninjured.

While it is desirable to do so, it is not practicable to search every building and shop along a dignitary's route. Doing so would also be a clear tip-off that a target worth attacking was going to be along shortly. Knowing that a worthwhile target will be along sooner or later is the key to any ambush.

AMBUSH OF DEFENSE MINISTER (BEIRUT, LEBANON, MARCH 20, 1991)

A car bomb aimed at Defense Minister Michel Murr killed about 10 people and wounded 38 on March 20, 1991. The bomb, made up of about 132 pounds of explosives cached in a Mercedes, was set off by a remote control as Murr's caravan of cars passed by. More than two dozen other cars were destroyed, as were several shops in the predominantly Christian Antelias community of Beirut. Murr was injured when the force of the blast blew his car over. The Defense Minister was en route from his home to a cabinet meeting in the largely Muslim western section of the capital when the motorcade was ambushed.

The calling of cabinet meetings in Lebanon at this period meant that various leaders had to travel from their redoubts in ethnic and religiously diverse sections of the city. Because of the security situation and crossing points only a limited number of routes of travel were available to officials. Car bomb ambushes were easy to set up along such routes, particularly when it was clear that targets would have to move along the streets to attend meetings.

Vehicle ambushes come in a variety of types. Attackers sometimes use vehicles to pace their targets, for instance.

AMBUSH OF U.S. LT. COL. JAMES ROWE (MANILA, PHILIPPINES, APRIL 21, 1989)

Attackers cut down U.S. Army Lt. Col. James N. Rowe and wounded his Filipino driver in an attack in a Manila suburb. The fatal ambush occurred as the officer was driving to work at Joint U.S. Military Assistance Group (JUSMAG) headquarters. The April 21, 1989, attack occurred about two blocks from the MAG building in Quezon City.

A car with three or four people inside pulled alongside Rowe's armor-plated vehicle in a traffic circle. Gunmen in the car (at least one of them armed with a .45 handgun and the other using an M16) fired at the vehicle at least 21 times in a chase of several blocks. They hit Rowe in the back of the head. The wounded driver got the car to military headquarters, where the fatally wounded Rowe was removed from the vehicle and taken to a hospital.

The attackers' vehicle was abandoned about four miles from the ambush site. The attack was later claimed by the New People's Army, which claimed Rowe was directly involved in counterinsurgency operations against the NPA. The NPA statement referred to "the firm commitment of the revolutionary forces to continue military action against U.S. personnel." In the aftermath of the attack, U.S. soldiers were advised to wear civilian attire to make them less conspicuous off base and were told to travel in groups and avoid darkened areas, such as alleys.

The salient points here were that no one picked up on the extensive preattack surveillance that went into this hit. The armored car wasn't armored to the need, either. The move to get U.S. military men into mufti outside the bases, and make them slightly less-conspicuous targets, was noteworthy for the lack of understanding about ambushes that it showed.

Rowe had been targeted as an individual, and the best

way to prevent that ambush was to pick up on it during the preparations. He wasn't targeted because he was just another guy in a uniform.

Besides, uniforms didn't make that much difference. Americans tend to stand out overseas, just as Japanese tourists are readily identifiable in the United States or Canada. Ambushers don't necessarily look for uniforms. They look for a particular type of target—and sometimes that target is a civilian.

SENDERO LUMINOSO AMBUSH OF EXECUTIVE (LA MOLINA, PERU, JULY 20, 1990)

The head of Peru's subsidiary of the B.F. Goodrich Company, his driver, and two of his bodyguards were killed in an ambush on July 20, 1990. Attackers, believed to be members of Sendero Luminoso (Shining Path), moved a tanker truck into a blocking position across a street in the wealthy capital suburb of La Molina and halted the two-car convoy of Antonio Rosales Durand. Rosales and the three people with him were killed in a barrage of gunfire and bombs. He was the chairman of Lima Caucho, SA, and was on his way to the company facilities when he was attacked.

NEW PEOPLE'S ARMY DOUBLE AMBUSH (THE PHILIPPINES, SEPTEMBER 26, 1989)

On September 26, 1989, two American civilians employed by Ford Aerospace and Communications Corporation at a military communications facility at Camp O'Donnell were killed by elements of the Communist New People's Army (NPA). The killings were timed to coincide with the arrival of U.S. Vice President Dan Quayle in the Philippines.

The NPA set up the afternoon ambush on a road near Capas. The men were killed after a dump truck and a jeepney blocked the highway between Capas and Camp O'Donnell. Half a dozen men popped up and opened fire at the car with sustained bursts of automatic-weapons fire. The bursts killed the Americans where they sat. But the attackers, wishing to make certain of their kills, opened the doors of the vehicle and poured in more rounds. The attackers fled in a second vehicle and left the dump truck behind.

That same day an officer in the presidential palace guard was assassinated in an attack about a mile from the presidential home. He was fatally shot as he drove in his car.

The dump truck across the road is generally an effective blockade, and it proved so in this case. There are few vehicles short of a tank that can push a dump truck out of the way no matter how expert the driver behind the wheel of the ambushed vehicle. With no place to go, surprise complete, and no weapons (with the possible exception of their own vehicle as a weapon), the outcome in this attack was never really in doubt. The ambushers got the men in the killing zone and carried out their attack carefully, even making certain the men were dead by moving in and firing at close range.

The second attack is largely notable because it demonstrates the difficulties many people face. They are not specifically high-profile targets; therefore, they don't look for tell-tale signs of an impending attack, such as surveillance. Unaware that they are being targeted—except in the most general way—they are watching for red lights, not watching for ambushers.

PALESTINIAN CAR AMBUSH OF ISRAELIS (GAZA STRIP, DECEMBER 7, 1992)

On December 7, 1992, Palestinian gunmen shot and killed three Israeli soldiers, firing from their car at an army vehicle near Gaza City. The soldiers, all reservists, were patrolling along a main road about 500 yards from the army's Nahal Oz roadblock at about 5:30 A.M. when they were ambushed. The road was busy with Palestinian laborers heading to Israel in the hours before dawn. In the darkness, three Palestinians in a white car overtook the soldiers and sprayed their jeep with automatic weapons fire, then fled. After the jeep was raked with fire, it crashed into a barrier blocking the entrance to a local neighborhood.

The Islamic Resistance Movement Hamas claimed responsibility for the Gaza City suburb attack in leaflets strewn nearby. The leaflets, issued by the Kassem military wing of Hamas, said the attack commemorated the uprising anniversary, the founding of Hamas on December 14, 1987, and the deaths of three Hamas activists killed by soldiers the previous week.

The fact is that ambushers, riding in cars passing by, are difficult to see until it is too late to take any effective action. Motorcycles are more easily maneuvered in traffic and, because of their high speed and small size, are often even more easily overlooked.

CYCLE-MOUNTED ATTACK ON JUDGE (BOGATA, COLOMBIA, AUGUST 16, 1989)

A quartet of cycle-mounted assassins gunned down Superior Court Judge Carlos Valencia in Bogota. Valencia had just confirmed warrants for Medellin drug godfather Pablo Escobar in connection with the 1986 murder of a newspaper publisher.

The attackers, who ambushed his vehicle about 10 blocks from his court, thwarted Valencia's body armor by shooting him in the neck area and through the armhole area of the armor. Three bodyguards with him were wounded in the August 16, 1989, ambush.

Even bodyguards aren't much help if they're not scanning the road and nearby areas for signs of problems at all times. The attackers had established enough information about the judge from surveillance to allow them to carry out the attack as he followed his regular routine.

AMBUSH OF POLICE OFFICIAL (MEDELLIN, COLOMBIA, AUGUST 18, 1989)

The murder of Judge Valencia sparked angry protests by judges. Security agencies scrambled to give judges the protection they demanded. Two days after Valencia's slaying, a national police colonel in Medellin, who normally had a bodyguard, redeployed those personnel to the court system. Waldemar Franklin Quintero was then ambushed and assassinated near his home. Attackers surrounded his car and stitched the vehicle with shots for several minutes. His slaying was claimed by drug-related attackers.

Redeploying his own guards was unwise. Good intelligence work by the drug lords' employees showed how vulnerable he was—and proved that the police did not really control the country. The attack also proved that bodyguards aren't any good if you don't have them with you.

NAGDI ASSASSINATION AMBUSH (ROME, ITALY, MARCH 16, 1993)

Two gunmen shot dead an Iranian opposition figure in the streets of Rome on March 16, 1993. Mohammad Hussein Nagdi, the representative in Italy of the National Council of Resistance of Iran, was shot twice as he went to work from his home nearby. Nagdi was ambushed by two men riding a Vespa scooter on a busy street. According to police, one of the men fired at his face and neck. The attackers then vanished in rush-hour traffic. Nagdi died in a police car on his way to a hospital.

Nagdi knew he was on a "hit list." He went to work carrying a gun, but he did not get a chance to use it. Nagdi was given police protection after the 1990 assassination of another opposition official in Switzerland, and although police were posted at his house and office, there was apparently none on the 100-yard route between them. Nagdi had just left his home for his workplace when the assailant fired. Morning. The victim was on the way to work. No security men were on the scene. Cycle-mounted gunmen race by. That was a recipe for death. There are many such recipes that a thoughtful ambusher can whip up.

AMBUSH OF PRESIDENTIAL CANDIDATE (SOACHE, COLOMBIA, AUGUST 18, 1989)

Front-running presidential candidate Luis Carlos Galan Sarmiento, a man opposed to and hated by the Medellin cartel, was shot down by narco-terrorist gunmen at a campaign rally in Soache. A local politico was also killed and more than half a dozen other people were wounded in the attack by no less than seven pistoleros waiting in a crowd. A bulletproof vest deflected some of the rounds, but Galan was killed by one of them.

Galan had promised to take firm action against the drug traffickers, and, as a result, they had put a $500,000 price on his head. He was a target, and he knew it. He had his guards, but the drug lords could buy more firepower than he could. They took advantage of a candidate's weakness: the need to attend public functions and "press the flesh." The ambushers used the crowd as cover and concealment as effectively as the VC used trees and jungle in Vietnam.

The killing of Galan sparked a collision between the drug lords and the government. Each side tried to prove the point that it ran the country. At least that ambush had a point. Sometimes the ambushes are pointless.

AMBUSH OF NUNS AND BISHOP (PUERTO CABEZAS, NICARAGUA, JANUARY 1, 1990)

Two Catholic nuns from the Order of Saint Agnes, one of them an American citizen and the other a Nicaraguan, were killed when their pickup truck was caught in an ambush near Puerto Cabezas, Zelaya, on January 1, 1990. Two others in the truck, a third nun and an auxiliary bishop, were wounded in the nighttime attack on the truck, which was marked with yellow crosses.

A rocket-propelled grenade peppered the vehicle. Some AK-47 rounds were fired at the truck as well. Sandinistas and Contra forces later traded potshots over responsibility for the gunshots, each side accusing the other of involvement in the attack and charging that the ambush was premeditated. Contra spokesmen flatly denied that the rebels had any units in the area where the attack took place; however, the area was known for being a Contra/Miskito rebel stronghold.

In all probability, it was an ambush of mistaken identity. Rebels assumed that anything moving after dark was hostile. The yellow crosses were invisible in the dark. To an extent, the victims "asked for it." The ambushed group had insisted on traveling after dark, though a group of religious representatives in another vehicle that was traveling with the victims had stopped earlier for safety's sake. The truck was also reported to have been recently purchased. It, and the sound of its engine, was totally unfamiliar to insurgents operating in the area. Mistakes are easy to make when conducting ambushes.

MISTAKEN IRA AMBUSH (COUNTY TYRONE, NORTHERN IRELAND, OCTOBER 4, 1992)

A Catholic man was seriously injured when his car was ambushed by gunmen in County Tyrone on October 4, 1992. A police spokesman said security forces believed the gunmen had mistaken the car for a police vehicle. The injured man, a passenger in the front seat, and the driver had no connection with the security forces. The shooting happened in the mostly Catholic County Tyrone town of Castle Derg, on the border of the Irish Republic and Northern Ireland. A primed mortar bomb of about 14 pounds was later found close to the scene, the Royal Ulster Constabulary (RUC) said, leading police to believe the ambush may have been a bungled IRA operation. The mortar bomb was not used in the attack.

It appeared the ambushers figured out their mistake fairly early and quit the attack, but this was a sterling example of fuzzy planning and poor execution in an ambush. Effective ambushes don't "just happen." And even the best-trained and most expert forces can muff an ambush.

FAILED ISRAELI AMBUSH (SOUTHERN LEBANON, APRIL 16, 1993)

Two women and a man were killed on April 16, 1993, in what appeared to be a bungled operation by Israel. Samir Sweidan, a Lebanese officer in the Popular Front for the Liberation of Palestine, was wounded when antitank rockets destroyed his car during an ambush on a road near Yater. Sweidan's wife was killed, along with a man and a woman who worked in his family's tobacco fields. Two Lebanese women, including his sister, were wounded.

The Israelis were apparently intent on abducting Sweidan when they ambushed his car with small-arms fire as he was taking the workers home. But Sweidan was not hurt at first. He dashed away in his car, speeding along the dirt road, when the rockets struck. The car exploded when it was hit by the rockets, and the wreckage was raked with gunfire.

But the unexpected arrival of Nepalese troops made the Israelis abandon their plans to capture Sweidan. The Nepalese UN soldiers had been driving from a nearby post in a truck when they saw the car explode. Though the truck had been damaged by shrapnel from the rockets, the soldiers drove on to break up the attack. At this the attackers then left the area on foot, with Israeli helicopters in the air in support. The ground team's escape was covered by the Israeli helicopter gunships.

In this case the Israelis clearly had done their homework. They knew where their target would be and had established a pattern. But the initial shots failed to halt the vehicle and their target started driving out of the killing zone. The attempt to take him out with a rocket failed, though it caused other casualties. The unexpected appearance of the Nepalese forces forced the attackers to break off contact and retreat without the injured target.

Not all cases are like this one. Not everyone just rolls up to an ambush or walks into it of their own accord, while carrying out their own routine. In some cases, surveillance is never carried out. Very little surveillance, except of the killing zone, is needed when the victim is summoned to the fatal rendezvous. Ruses are sometimes used to entice the victims in before the trap is sprung.

MNLF AND COMMUNIST "SURRENDER" AMBUSHES (THE PHILIPPINES, 1990–1993)

On January 3, 1990, about 130 members of the Moro National Liberation Front (MNLF) promised to surrender near Buldon, Maguindanao, but instead attacked trucks and government agents coming to accept the capitulation. Three people were killed in the ambush, but a soldier survived and reported the events.

Then on July 30, 1992, seven soldiers and two civilian negotiators were killed in an ambush while trying to persuade a communist rebel leader in the southern Philippines to surrender. Some of the rebels opened fire during talks held to persuade the rebel commander, identified as Commander Jabbar, and 60 of his men to surrender. Four other members of the military Special Action Force were reported captured by the rebels, who became angry after their terms for surrender were rejected by the army officer negotiating the surrender. The government force had been called out to talk about the possible surrender and was not expecting trouble.

Rebels ambushed and killed four Philippine soldiers seeking the surrender of a communist guerrilla on southern Mindanao Island on April 25, 1993. Two other soldiers were wounded in the attack in Bakalan municipality, Zamboanga del Sur Province. The soldiers, led by a young lieutenant, went to Bakalan to negotiate the surrender of a New People's Army guerrilla, but the rebel and an intermediary who arranged the meeting did not show up. The soldiers were ambushed as they were returning to base.

The same tactic, it seems, worked over and over in the Philippines. The job of the military was to eliminate armed opposition by any means necessary, including the acceptance of surrenders. That made the soldiers vulnerable over and over again. Notably, however, the ambushed parties were heavily outgunned and outmanned by the attackers. In a very real sense, they invited trouble by going to the "surrender meetings" with a force too small to react to, and resist, an ambush.

Soldiers are not the only ones who have a job to do. Anyone can be lured to his or her doom if given what seem to be sufficiently sound reasons.

AMBUSH OF CAB DRIVER (BELFAST, NORTHERN IRELAND, APRIL 17, 1991)

A Roman Catholic cab driver was ambushed and slain in southern Belfast on April 17, 1991. He was lured to the ambush by a phone call, then shot down by a trio of attackers. The Ulster Freedom Fighters, a Loyalist paramilitary organization opposed to the IRA, claimed the attack. They said the IRA employed taxi drivers in targeting Protestants.

Lures are not the only type of ruse that is effective in an ambush.

CHRISTMAS DAY AMBUSH (WELI OYA, SRI LANKA, DECEMBER 25, 1992)

On December 25, 1992, Liberation Tigers of Tamil Eelam (LTTE) rebels ambushed government troopers in the Weli Oya area, killing 42 soldiers. The rebels used remote-control mines, rocket grenades, and small arms in their assault against three platoons returning to their bases. To gain the element of surprise, the rebels, about 250 of them, dressed in military-style uniforms. And, in order to delay the dispatch of reinforcements for the government troopers, the rebels staged simultaneous attacks against two nearby army bases. Despite that move to split the government forces and give the ambushers more time to escape, artillery and helicopter gunships responded to the ambush. The attackers lost up to 30 of their own effectives, including an LTTE major who was reported killed in the two-hour battle.

Using government uniforms gave the rebels an initial advantage in this carefully planned ambush. The effort to delay reinforcement of the ambushed soldiers and to gain more time to escape by having other elements attack nearby bases was masterful. But the rebel commander waited far too long and spent too much time fighting once the element of surprise was lost. Artillery travels faster than aircraft, though helicopter gunships are often more deadly when they arrive on the scene. The losses were far too even to make this ambush a rebel victory. In fact, most would call it a strategic loss for the LTTE, despite the tactical victory the rebels achieved in the first moments of the encounter. The rebel commander snatched defeat from the jaws of victory by waiting too long to pull back.

The government troops were not expecting soldiers in the open, wearing their own uniforms, to open fire. They expected fire to come from other spots. So, too, soldiers expect ambushes to come from behind fences, from public spots. They don't expect deadly ambushes will be mounted from private homes.

IRA ATTACK ON ARMORED POLICE VEHICLE (BELFAST, NORTHERN IRELAND, MAY 1, 1991)

A policeman was mortally wounded and three others were injured on May 1, 1991, when the IRA ambushed an armored police vehicle in Belfast. The night before the attack, an IRA ambush team took over two homes for use in the rocket and gun attack.

The IRA has a long history of this type of activity. They seize a house or car needed in an attack about 12 hours before the attack is staged. In most cases the family who has the car or house is held hostage until after the rebel attack or ambush is carried out. Then the hostages are freed.

It is not unusual for ambushers in some places to hold a family, or even an entire village, hostage.

SHINING PATH AMBUSH OF MILITARY PATROL (HUINGE, PERU, APRIL 7, 1993)

About 50 Shining Path guerrillas ambushed a military patrol of 30 Peruvian soldiers near the jungle city of Tarapoto, 370 miles north of Lima near the Alto Biabo area. The patrol, after crossing rapids of the Huallaga River on April 7, 1993, entered the village of Huinge where guerrillas attacked; they fired rifles and detonated explosives from both sides of the road.

During the battle, Alto Biabo area governor Hector Lopez was killed along with seven soldiers. Another 10 soldiers were wounded. At least 10 rebels were also killed in the shoot-out, though the bodies were removed by their comrades and dragged into the jungle, police said. The military patrol was on its way to reinforce a base that had been repeatedly attacked by about 180 guerrillas believed active in the area.

Soldiers, not unreasonably, see a village and habitation as a "safe area" when they have spent hours slogging along through underbrush and along roads where every birdcall can be a signal for attack. The rebels capitalized on the human reactions of the troops and caught them from both sides of the road as they entered the village. It is worthwhile noting that the troops were attacked as they were en route to reinforce other soldiers who had suffered attacks. A common one-two punch used by rebels is to attack a fixed position or ambush a group of soldiers, then attack the relief forces or reinforcements.

A lack of air cover, as well as the failure to locate the explosive charges laid along the road, contributed to the rebels' success in this case. Air cover is a scarce resource, but one that can be as invaluable as it is costly to use. But air cover is not a panacea. There are some ambushes that an aircap can never affect, negatively or positively. If there is a single lesson that has to be learned about ambushes it is to always expect the unexpected and to do the unlikely. Not all ambushes require arcs of fire or lay-up points. Some are as simple as they are unexpected—and are equally deadly.

MILF POISON AMBUSH (BANISILAN, PHILIPPINES, DECEMBER 1993)

In early December 1993, three Filipino soldiers died and 25—including a battalion commander—became sick after drinking from a well that was apparently poisoned by fleeing Moslem rebels. The soldiers had run out of water when they drank from the well while pursuing a group of guerrillas in Banisilan municipality, Mindanao, in the southern Philippines. The well was poisoned by rebels of the Moro Islamic Liberation Front (MILF), one of three rebel factions seeking Muslim self-rule.

Ambushes are indeed the weapon of the weak, an "equalizer" in every respect.

SNIPER AMBUSH (SOUTH ARMAGH, N. IRELAND, FEBRUARY 25, 1993)

On February 25, 1993, a sniper using thick cover in Ulster's border area killed a policeman near the Irish border. The policeman was killed while accompanying a British army patrol in a field off Castleblaney Road in the town of Crossmaglen, South Armagh. The area was widely recognized as "bandit country."

The constable was killed by a single shot as he patrolled a field alongside British soldiers in an area of a rolling bog. The bullet sliced through the constable's flak jacket, police said, suggesting that the sniper used a long-range, high-powered specialist rifle believed to be in the IRA arsenal.

The IRA snipers struck again on St. Patrick's Day, Ireland's National Day, killing a member of a patrol near the border village of Forkhill. Within a matter of about six months, two soldiers and two policemen had been shot dead by the long-range sniper, or snipers, in South Armagh.

Single shots were being used throughout the series of sniper ambushes. The IRA reasoned that troops, alerted by the sound of the first round, would not be able to pinpoint the location of the marksman in the dense underbrush by observing the point of origin of a following round. The number of reports of single rounds being fired, rounds that failed to hit anyone, showed the sniper was no superman behind his sights.

But his, or their, attacks had troopers and officers alike on edge.

Security chiefs as well as soldiers and policemen on patrol grew concerned at the sniper's use of the U.S.-made Barrett "Light Fifty" weapon, which has a range of up to 1,800 yards and fires heavy machine-gun cartridges. The sniper killings prompted press speculation that the IRA was using either a U.S. mercenary or an ex-British soldier who had been "turned." Press attention and the continuing carnage caused by the sniper led the British army commander in Northern Ireland to announce that hunting the IRA sniper who had been picking off his soldiers was a top priority.

Extra measures were then taken to safeguard patrols against sniper fire, particularly using air cover. These sniper attacks diverted large amounts of men, and scarce resources, from the larger campaign in Northern Ireland. Instead of tackling the IRA infrastructure and larger units, the manpower and flight hours were used in an effort to track down the ambush menace behind the telescopic sights.

AMBUSH OF POLICE CHIEF (TIJUANA, MEXICO, APRIL 28, 1994)

The outspoken corruption-busting police chief of Tijuana, Mexico, was gunned down in a nighttime ambush by unidentified gunmen carrying assault weapons. The gunmen killed Frederico Benitez and his bodyguard/driver as they drove through the outskirts of the border city, using a blocking car to cut him off and then lacing Benitez' vehicle with automatic-weapons fire.

Benitez was an anticorruption crusader who in 16 months had ousted well over a third of the officers from the 2,100-member municipal police force. His bodyguard and driver, an 18-year police veteran, was killed instantly. Benitez was shot twice, with one bullet piercing his neck and skull, and died later in a hospital. The police chief had been summoned into town by an urgent call about a bomb threat to municipal buildings—a threat that proved to be false. Benitez and his bodyguard were shot and killed as they were driving back to the chief's house.

The ambush took place near the neighborhood where presidential candidate Luis Donaldo Colosio was assassinated just weeks before. Benitez had claimed that the Colosio attack was carried out by more than one man—a thery that did not jibe with Mexico City's version of events.

This ambush was carefully crafted: a false report to bring the police chief to the scene and a strike on the way home, when the released tensions of the phony report made both the driver and chief less cautious.

• • •

Such is the power of the ambush. Ambushes preserve the personnel and assets of the ambushers. They make war last longer. And if the ambushers can make a war last long enough, they will destroy the will of their opponents—even if they cannot crush the military capability of that enemy. That's winning through attrition. It may not be gentlemanly. The purists might call it cowardly. But it is smart. It is effective. And it has millennia of proven success behind it.

BIBLIOGRAPHY

This book was written in part because there was a paucity of material on the subject of ambushes; nonetheless, there are some specific books and data bases worth reviewing.

Ambush and Counter Ambush. Boulder, CO: Paladin Press, 1988.

The Art of War. Sun-tzu. Translated by Thomas Cleary. Boston: Shambhala Publications, 1991.

Combat Training of the Individual Soldier and Patrolling, FM 21-75. Washington, D.C.: Department of the Army, 1962.

Guerrilla. Charles W. Thayer. New York: Signet Books, 1965.

The Guerrilla and How to Fight Him: Selections from the Marine Corps Gazette. Edited by Lt. Col. T.N. Greene. New York: Frederick A. Praeger, Inc., 1962.

Guerrilla Warfare and Special Forces Operations, FM 31-21. Washington, D.C.: Department of the Army, 1961.

International Security Database. San Diego, CA: Vantage Systems Inc., 1994.

Land Mine Warfare, FM 20-32. Washington, D.C.: Department of the Army, 1963.

Lessons Learned Number 15—Ambushes. San Francisco, CA: MAAG, 1963.

Lessons Learned Number 27—Ambushes. San Francisco, CA: MAAG, 1963.

Living in Troubled Lands. Patrick Collins. Boulder, CO: Paladin Press, 1981.

Low Intensity Operations—Subversion, Insurgency & Peacekeeping. Frank Kitson. London: Faber & Faber, 1975.

Manual of the Mercenary Soldier. Paul Balor. Boulder, CO. Paladin Press; 1988.

Mao Tse-tung on Guerrilla Warfare. Translated and with an introduction by Brig. Gen. Samuel B. Griffith. New York: Frederick A. Praeger, Inc., 1967.

Mata Handbook. Fort Bragg, NC.: The United States Army Special Warfare School, 1964.

Modern Guerrilla Warfare—Fighting Communist Guerrilla Movements. Edited by Franklin Mark Osanka. New York: The Free Press of Glencoe, a Division of the Macmillan Company, 1962.

NSW/YSMC Riverine Operations Handbook XL-00080-01-93. Strategy and Tactics Group. San Diego Naval Special Warfare Center, 1993.

Operations Against Guerrilla Forces, FM 31-20. Washington, D.C.: Department of the Army, 1951.

Professional Knowledge Gained from Operational Experience in Vietnam, NAVMC 2614. Washington, D.C.: Department of the Navy, 1967.

Ranger Training, FM 21-50. Washington, D.C.: Department of the Army, 1957.

"Terrorist Attacks on Vehicles." Morris Grodsky. *Assets Protection Magazine*, 1981.

Terrorist Group Profiles. Washington, D.C.: U.S. GPO, 1988.

Transportation Security. Tony Scotti. Somerville, MA: Scotti School of Defensive Driving, 1979.

Unconventional Warfare Devices and Techniques TM 31-200-1. Washington, D.C.: Department of the Army, 1966.

U.S. Army Counterinsurgency Forces, FM 31-22. Washington, D.C.: Department of the Army; 1963.

Bibliography

Database of World Political Violence. San Diego, CA and Washington, D.C.: Vantage Systems Inc., 1994.

War in the Shadows: The Guerrilla in History. Robert B. Asprey. Garden City, NY: Doubleday & Company, Inc., 1975.

ABOUT THE AUTHORS

Gary Stubblefield, LCDR USN (Ret.), is one of the original SEAL "Men in Green Faces" of Vietnam. Stubblefield, who commanded SEAL Team 3 during his career, knows ambush and counterambush tactics from extensive personal experience. A co-owner and vice president of Vantage Systems Inc., an international security management firm, Stubblefield is now active in protecting Americans overseas as well as teaching executives to protect themselves in hostile environments.

Gary Stubblefield (left) with Mark Monday (right).

Mark Monday is an award-winning writer who specializes in insurgency and revolt. With more than 25 years' experience in professional journalism, he was formerly editor and publisher of both *Terrorism, Violence and Insurgency (TVI) Journal* and *Briefing: Terrorism and Low-Intensity Conflict.* The director of information services for Vantage Systems Inc., an international security management firm, he shares a Pulitzer Prize.